Maritime Clusters and the Ocean Economy

Over the last decade the concept of clustering has become a central idea for analyzing the competitiveness of nations, industries and firms. This book shows how the cluster concept can be usefully applied to the study of maritime activities.

Such activities, including shipping, shipbuilding and port and maritime services, are clearly geographically concentrated in a number of maritime clusters. However, as the author shows, these are having to compete with other uses of the coasts and oceans including capture fisheries, marine aquaculture, offshore energy and tourism. Sound governance and planning is therefore required to manage the competing claims for ocean space. The book shows how competing industries and other stakeholders can cooperate and benefit from an integrated approach to the development of maritime clusters.

The contribution of approaches such as integrated coastal zone management and innovations such as ocean business councils, as well as coordinated networks of maritime clusters are reviewed. Case studies are included from around the world, including detailed examples of the development of the Nelson Mandela Bay Maritime Cluster in South Africa and from Poland in the Baltic Sea.

Peter B. Myles is an International Tourism Consultant, and Owner and Founder of Tournet Africa, based in Port Elizabeth, South Africa. He was responsible for developing the first module on coastal and marine tourism in Africa included in the postgraduate BCom (Tourism) Honours program offered at the Nelson Mandela Metropolitan University, South Africa. He is currently Chairman of the Nelson Mandela Bay Maritime Cluster.

Earthscan Oceans

Marine and Coastal Resource Management
Principles and Practice
Edited by David R. Green and Jefferey Payne

Marine Transboundary Conservation and Protected Areas
Edited by Peter Mackelworth

Transboundary Marine Spatial Planning and International Law
Edited by S.M. Daud Hassan, Tuomas Kuokkanen, Niko Soininen

Marine Biodiversity, Climatic Variability and Global Change
By Grégory Beaugrand

Marine Biodiversity Conservation
A Practical Approach
By Keith Hiscock

The Great Barrier Reef
An Environmental History
By Ben Daley

Governing Marine Protected Areas
Resilience Through Diversity
By Peter J.S. Jones

Marine Policy
An Introduction to Governance and International Law of the Oceans
By Mark Zacharias

For further details please visit the series page on the Routledge website:
www.routledge.com/books/series/ECOCE

Maritime Clusters and the Ocean Economy

An Integrated Approach to Managing Coastal and Marine Space

Peter B. Myles

Routledge
Taylor & Francis Group
LONDON AND NEW YORK

earthscan
from Routledge

First published 2017
by Routledge

2 Park Square, Milton Park, Abingdon, Oxfordshire OX14 4RN
52 Vanderbilt Avenue, New York, NY 10017

Routledge is an imprint of the Taylor & Francis Group, an informa business

First issued in paperback 2019

British Library Cataloguing-in-Publication Data
A catalogue record for this book is available from the British Library

Library of Congress Cataloging-in-Publication Data
Names: Myles, P. B., author.
Title: Maritime clusters and the ocean economy : an integrated approach to managing coastal and marine space / Peter B. Myles.
Description: New York : Routledge, 2017. | Series: Earthscan oceans | Includes bibliographical references and index.
Identifiers: LCCN 2017011436| ISBN 9781138182882 (hardback) | ISBN 9781315646152 (ebook)
Subjects: LCSH: Marine resources. | Shipping—Economic aspects. | Industrial clusters.
Classification: LCC HC92 .M95 2017 | DDC 333.91/64—dc23
LC record available at https://lccn.loc.gov/2017011436

ISBN: 978-1-138-18288-2 (hbk)
ISBN: 978-0-367-33536-6 (pbk)

Typeset in Goudy
by Keystroke, Neville Lodge, Tettenhall, Wolverhampton

Contents

Illustrations

Figures

Tables

Preface

I was born in Ndola, Northern Rhodesia, before the country received its independence and was renamed Zambia. On numerous occasions when invited to speak at international conferences on marine tourism and coastal recreation, and after the program director had introduced the speakers, I am asked the same question, 'How is it that someone like yourself born in a landlocked central African country has such an obvious love of the sea and passion for responsible ocean governance?' Perhaps the next few pages will explain how one thing has led to another and the combination of empirical observations and practical experiences over a period of time have brought me to where I am now.

It all began with a love of water. The first ten years of my life were spent in Livingstone, a city with a hot climate near the mighty Zambezi River. As kids we spent most of our time cooling off in the local swimming pool or in the Zambezi River. We knew where it was safe to swim in the river by recognizing the signs associated with a presence of crocodiles and hippos. Most of our social life in those days was spent around the local municipality swimming pool. This was also a great place for boys to meet girls. As a result of our natural affinity to water we became good swimmers and also good underwater swimmers. We would challenge each other as to who could swim the most lengths underwater holding our breath in an Olympic size swimming pool. We would practice holding our breath for as long as possible using a wristwatch for timing ourselves. I developed my lung capacity until I could swim three lengths underwater and could hold my breath for about three minutes. At about this time when we were experimenting with land-based underwater experiences the *Popular Mechanics* magazine was publishing a series of articles on how to make your own rubber mask and flippers. This coincided with the first books published on underwater swimming written by the early pioneers such as Hans Hass and Jacques Cousteau. The first books were followed by the first underwater films such as *Under the Red Sea* and *The Silent World*. We were fascinated by the natural beauty of this relatively unexplored world in the 1940s and 50s.

The books and films added a new dimension to our land-based underwater experiences and we began swimming underwater in dams, lakes and other large expanses of water such as quarries where the visibility was clear enough to see below the surface. We soon learned that what you see on the surface has very little bearing on what is below the surface. We found all kinds of junk like old tyres, rusty pots and pans, building rubble, and even the odd skeleton underwater. Little did I know at that time that one day I would have the same experience when either snorkelling or scuba diving at seaside resorts and in the different oceans.

In those early colonial days on the continent of Africa most of us were sent to boarding school. I was fortunate that my father sent me to a high school in the Western Cape province of South Africa where I spent as much time as possible snorkelling using my lung capacity developed in swimming pools to explore the beautiful coastline so rich in colourful coral reefs, abundant sea life and multi-coloured fish species. Such was the perfect ecological balance in those days that we could dive in at Clovelly and swim across Fish Hoek Bay to Sunny Cove. The thought of being taken by a Great White Shark never entered our minds. Today no person in their right mind would swim across Fish Hoek Bay after a number of fatal shark attacks.

Something else that I learnt about swimming pools is that they have to be maintained. We did this by testing the pH of the water and then adding either chlorine or acid to maintain the correct balance for perfect visibility. During a child's birthday party on a hot summer day the kids spent most of the time in the pool. The following day I could not believe my eyes because the water was dark green with poor visibility. No matter how hard I tried I could not get the water back to the correct pH balance by adding either acid or chlorine. I came to the conclusion that the kids must have all urinated in the pool and eventually I had no choice but to empty the pool and start from scratch. At a later stage I remembered this experience and a random thought crossed my mind. Could the ocean, like a huge swimming pool, become so polluted that it would also change colour? The difference is that we cannot empty the ocean and start again.

During my years as a tourism specialist I observed on many occasions that institutional structures were not working. National, regional and local tourism organizations were totally dependent on government funding and were subject to a high level of instability. Any change in political structures or government policy usually meant a change in institutional support. In fact unless tourism structures were partly funded by the private sector, usually by the introduction of a transient room tax or bed levy, then local tourism organizations were at the mercy of the local authority. The message was quite clear, 'Don't bite the hand that feeds you' and hence public funds could be used to manipulate strategies that would benefit local authorities and not necessarily benefit the tourism stakeholders and product owners.

In the 1980s the International Association of Convention and Visitor Bureaus (IACVB) was perhaps the first global tourism institution to recognize that there were benefits in forging mutually beneficial cooperative alliances. This was the beginning of understanding and appreciating that stakeholders in the same industry could be independent and competitive but were in fact totally interdependent on common infrastructure such as roads, rail and air transport.

Hence, no single product owner actually owns the whole product. The product is a combination of different accommodation establishments, attractions, events and activities, which all add up to an appealing tourism experience. A tourist route that is not developed and marketed collectively by the product owners along that route is likely to fail and will become either a virtual route or a highway to nowhere with direction signs along the route but no champion to coordinate all the interesting things to see and do along that route.

In conclusion, and taking into consideration all these different experiences encountered over a period of time, the author believes that collaborative economic development, commonly referred to as 'cluster development', stands out as the most promising management tool for responsible ocean governance. Land can be divided by various territorial barriers and boundaries such as walls and fences but the ocean is much more complicated. The demarcation of territorial boundaries in coastal zones and the deep ocean can only be managed effectively by cooperative agreements and alliances. Clusters do not happen naturally; they have to be constructed. Furthermore, there is no 'one size fits all' model. Over a period of time a cluster will develop its own size, shape and personality.

A cluster is defined as a geographic concentration of interconnected companies and institutions working in a common industry. The maritime industry is ideally suited to the development of maritime clusters. Maritime cluster organizations represent almost every traditional maritime sector, although in practice this does not always seem the case. Some maritime sector organizations answer to a broader definition of maritime sectors, such as navy and coastguard, inland navigation and maritime works. The fisheries sector, marine tourism and coastal recreation sectors are sometimes represented by the national cluster organizations, although less frequently than the traditional maritime sectors. There are different approaches to developing maritime clusters. This initiative can be taken top–down resulting in government induced organizations, or induced by leader firms and/or sector associations (bottom–up). The common objective with both approaches should include a strong focus on competitiveness. Opportunities to increase competitiveness come not from a single asset or an investment, but from the combination of people, place, investment and attitude. In the final analysis, clusters are not constrained by rigid structures, rules and regulations. Successful clusters are driven by passion and commitment.

Peter Myles, 2016

Acknowledgements

In compiling this book I have received encouragement and constructive criticism from a number of respected colleagues from many disciplines. They have contributed to my collective knowledge and experience over a period of 14 years, ever since I first entered the field of tourism specialisation now commonly referred to as Coastal & Marine Tourism or Coastal & Maritime Tourism depending on the emphasis. I cannot possibly name them all even though I would like to. However, I would like to recognise some key individuals who have helped greatly in many different ways as I have chartered my direction going forward from coastal and marine tourism education and awareness to responsible ocean governance in the blue economy. In no particular order of priority I acknowledge my respected colleagues who are also my valued friends.

Marc Miller, Adjunct Professor, School of Marine & Environmental Affairs, Department of Anthropology, at the University of Washington, Seattle, is internationally renowned for his work on marine natural resources, their use and the search for ways that they can be sustainably developed. As a cultural anthropologist, he looks not only at how people build their lives around the coasts and oceans through direct exploitation, but also at the other ways they value them aesthetically, ethically, and even spiritually. Together with Vlad Kaczynski, they are the pioneers who first introduced coastal & marine tourism (CMT) to South Africa, and in fact to Africa. I was privileged to work with both of them when we developed the first postgraduate module on CMT.

Vlad Kaczynski, Associate Professor (Ret), School of Marine & Environmental Affairs, at the University of Washington, Seattle, and Adjunct Associate Professor, Henry Jackson School of International Studies, at the University of Washington, has also served as a Visiting Professor at the Nelson Mandela Metropolitan University, Port Elizabeth, South Africa. He has devoted his academic and professional career to studies of the world ocean affairs, with special focus on the developing countries of Africa and South America and marine economies in transition. As a Senior Fulbright Scholar supported by the US Department of State, Kaczynski has served as a consultant for a number of international donor organizations.

Paul Holthus, founding President and Chief Executive Officer of the World Ocean Council, works with the private sector and market forces to develop practical solutions for achieving sustainable development and addresses environmental concerns, especially for marine areas and resources. His experience ranges from working with the global industry associations or directors of UN agencies to working with fishers in small island villages. He has been involved in coastal and marine resource sustainable development and conservation work in over 30 countries in Europe, Asia, the Pacific, Central America and Africa. As a consultant on sustainable development and environmental management, he has worked with companies, industry associations, UN agencies, international NGOs and foundations on sustainability, especially in the areas of oil/gas, fisheries, aquaculture, standards and certification.

Patrick H.G. Vrancken, Associate Professor at Nelson Mandela Metropolitan University (NMMU), is the current incumbent occupying the Chair in Law of the Sea and Development in Africa which is hosted by NMMU, funded by the Department of Science & Technology (DST) and managed by the National Research Foundation (NRF). Vrancken is the author of numerous research outputs in national and foreign accredited law journals and books. He is the editor of, and one of the contributors to, *Tourism and the Law in South Africa*, and also the author of *South Africa and the Law of the Sea*. More recently, he has played a major role in the development of the maritime training and research agenda both at national level and at the NMMU.

To my wife and family, thank you for your support and encouragement during the long hours of desktop research, writing and editing of the book, which left very little spare time for social interaction.

It is my wish that this book will encourage and inspire readers to explore the silent world of the deep blue oceans and aquamarine seas which cover just over 70% of our planet. Individually we may not be able to change the world but collectively we can make a huge difference in our sphere of influence (i.e. in our ocean space where we are stewards of our valuable coastal and marine resources). This calls for responsible ocean governance collectively by coastal authorities, the ocean business community, stakeholders, academics and recreational users.

Introduction

The ocean is an interconnected global ecosystem supporting a wide range of uses. Maintaining a healthy ocean requires responsible stewardship by all users. The ocean supports a growing range and level of commercial and other activities. User conflicts are on the rise, as are impacts from increasing use. Concerns about impacts may limit access to marine areas and resources. Ocean users, including those who simply rely on shipping to move goods, are increasingly being held accountable by governments and NGOs for the state of ocean health. Continued use will increasingly require a social licence to operate above and beyond simple regulatory compliance. Many companies are trying to reduce their marine ecosystem footprint. However, the best efforts by a single company or an entire industry will not be enough to address cumulative impacts in the interconnected marine ecosystem. Companies dependent on the ocean will benefit from collaboration with others in the ocean business community to address marine environmental issues.

Coastal resources cannot be used by any interest group as their exclusive right. The sea cannot be regarded as a common basin for effluent disposal. Almost all coastal and marine areas produce or support multiple products and services. Sectoral solutions usually 'transfer' the problem between resources, products and services. Tourism will not flourish if the area loses its attractions to visitors; and fisheries are usually on the receiving end of everyone else's problems. Industry and energy facilities can degrade the environment for all other activities. There is, therefore, a need to bring sectoral activities together to achieve a commonly acceptable coastal management framework. The challenge, which is a recurring theme throughout this book, is to develop a vehicle that will encourage collaboration and consultation at all levels within a responsible ocean governance domain.

This book investigates institutional arrangements such as maritime clusters that bring together the diverse ocean business community (shipping, fisheries, oil and gas, aquaculture, offshore renewable energy, tourism, marine technology, manufacturers, retailers, insurers, finance, etc.) to achieve the business benefits of collaboration in addressing shared marine environmental issues. Working groups such as the World Ocean Council (WOC) are developing practical solutions to a number of important marine issues, for example, marine spatial planning, marine invasive species, marine debris, ocean noise, marine mammal impacts, water pollution, improved ocean science, and ocean policy and governance. The UN Conference on Sustainable Development in June 2012 (Rio+20), perhaps more than any previous conference on sustainable development, highlighted support for the sustainable use and conservation of oceans to achieve a 'blue economy'. A blue economy is one in which ocean ecosystems bring economic and social benefits that are efficient, equitable and sustainable. Used wisely, ocean resources can help address poverty, food security, sustainable livelihoods and conservation.

The world's oceans are increasingly under pressure from rising sea temperatures, ocean acidification, overfishing, biodiversity loss, habitat loss and pollution. Key ecosystems such as coral reefs may soon reach critical thresholds, disproportionately impacting on the people and communities that are most vulnerable, such as small island states and coastal developing countries. Furthermore, the world's population is continuing to increase, and with it the demand for food from the oceans and elsewhere will increase. The solution lies in finding ways to move beyond the usual tradeoff between economy and the environment. Coastal and ocean stakeholders need to ensure that the ways in which they value, manage and use the oceans sustain the habitats and ecosystems that provide food and livelihoods. People-centred approaches to marine biodiversity conservation illustrate how this can be achieved. Coastal and ocean stakeholders recognise that sustainable incomes and livelihoods require healthy marine ecosystems; and equally, that healthy ecosystems can only be achieved if people and communities have sustainable incomes and livelihoods. Such 'win–wins' can be achieved when marine ecosystem considerations are an integral part of social and economic decision-making, rather than separate and sometimes competing matters. There are no 'one size fits all' actions that can deliver a blue economy. Some are global in scale (such as governance of the high seas), but many are needed at local, national and regional levels.

Managing the multiple uses of the ocean is therefore vital. Successful management of any activity, however, requires adequate understanding both of the activity and of the context in which it takes place. Such an understanding is needed even more when the management tasks are divided among many players: unless each knows how the part they have to play fits into the overall pattern, there are risks of confusion, contradictory actions and failures to act. Managing human uses of the ocean has inevitably to be divided among many players.

Manoj Gupta, author of *Indian Ocean Region: Maritime Regimes for Regional Cooperation* (Springer Science, 2010), pointed out that oceans, regarded as the 'last frontier' on earth, shall decisively influence world geopolitics, globalisation, socio-economic balance and life on earth. Today, 82% of all nations border the world's oceans or seas. Seaside holidays have a long history. They were popular for several hundred years among the ruling classes of the Roman Empire. Over the past few decades, coastal tourism has come to include creating opportunities for the public to enjoy the local wildlife. This has generated a large number of businesses serving tourists. However successful management of sustainable tourism requires a complex balancing exercise. It will require the involvement of a wide range of authorities, residents and commercial interests. At the same time, within any specific resort, there will be conflicting demands for marine space: among others, sunbathing versus beach volleyball, swimming versus jet-skis and similar devices, and wildlife watchers versus water skiers. All this will require management alongside the many other demands on coastal marine space from fishing (both commercial and recreational), shipping, ports, sand and gravel dredging, and all the other human activities that affect the coastal zone. There is an urgent need for a more integrated approach, at the global level as well as at the regional and sub-regional levels.

This book investigates and provides evidence that maritime clusters can contribute significantly to responsible ocean governance and the blue economy. However clusters do not happen naturally; they have to be constructed. Over the last decade, the concept of clustering has become a central concept for analyzing the competitiveness of nations, industries and firms. The cluster concept can be usefully applied to study clustering of maritime activities. Maritime activities, such as shipping, shipbuilding and port and maritime services, are clearly geographically concentrated in a number of maritime clusters. Due to ongoing internationalization in these industries, the concentration of maritime activities in clusters is likely to increase.

Based on empirical data from numerous surveys conducted amongst firms located in maritime clusters, studies have shown that the cluster structure is beneficial for improving maritime performance and competitiveness. This book concludes with recommendations and guidelines that strongly support a transition from the accepted definition of coastal and marine tourism to a definition that embraces maritime tourism, i.e. coastal and maritime tourism. This is a necessary step to ensure that marine tourism and coastal recreation are integrated into a responsible ocean governance framework. The framework will incorporate an integrated maritime strategy and policy that will ultimately create a more desirable state for tourists and a more desirable state for the marine environment in each and every local maritime domain. A 'who's who' of competitiveness practitioners from around the world gathered for The Competitiveness Institute's conference in Spain during October 2012 (see: www.tci-network.org/tci2012). Each participant received an excellent workbook with abstracts of the 200+ papers. A quotation from this workbook perhaps sums up the significant contribution of clusters to socio-economic development:

> Opportunities to increase competitiveness come not from a single asset or an investment, but from the combination of people, place, investment and attitude.
>
> Gerard Quinn, Grow Wellington, New Zealand

Based on research for convincing evidence that clusters can contribute significantly to responsible ocean governance and the blue economy, this book examines various examples of collaboration from onshore maritime clusters to intercontinental alliances in the deep ocean.

This book is a collection of many contributions by different authors and researchers over a period of time. Every attempt has been made to request permission to quote from original work and to acknowledge different authors and contributors. However, in some cases the author was unable to trace the original source of information or the contribution of writers who have since passed away. It is hoped that in the spirit of sharing information which could perhaps make a difference to informed decision making, that those who have not been acknowledged will accept that their work has made a difference and is making a difference.

In this regard the author has adopted the 'Spirit of Ubuntu' which acknowledges that 'I am who I am because of who you are'. Ubuntu is an Nguni Bantu term roughly translating to 'human kindness'. It is an idea from the Southern African region often translated as 'humanity towards others'. It is also used in a more philosophical sense to mean 'the belief in a universal bond of sharing that connects all humanity'. In Southern Africa, it has come to be used as a term for a kind of humanist philosophy, ethic or ideology. The author is the recipient of incredible generosity with respect to shared knowledge. Hence in the 'Spirit of Ubuntu' the author attempts to share with the readers of this book the knowledge gained from numerous authors and from different experiences.

Chapter 1

Collaborative economic development

Structure and strategy

The question of which comes first, structure or strategy, has been widely debated for many years. In business management it is commonly accepted that the successful implementation of strategy requires an appropriate structure. However, organizational structures have two characteristics similar to physical structures, i.e. they provide support and they are an immovable object. If the original support no longer serves a purpose then it has to be demolished to make way for a new structure. This can be a costly exercise with complex legal implications. The author hopes that the readers of this book will be able to make up their own mind which institutional arrangements for the implementation of effective ocean governance strategies and policies offer the best option going forward.

To try and fit all the actors and players associated with the oceans and seas into a local or regional maritime organizational structure would be a daunting task. Who or what organization would govern this structure? Would it be led by government and driven by the private sector or vice versa. Have we reached a point where the oceans and seas are now everyone's business? Users of the ocean's resources cover a wide spectrum of stakeholders from coastal recreationists, to sun worshippers and beach lovers, to boating enthusiasts, surfers and divers, and also to firms that depend on imports and exports of goods and services for survival.

Management work is the planning, organizing, leading, and controlling work that can be effectively done only by people in management positions. Managers should be selected for their demonstrated and potential ability to manage. They should have clearly defined management work they are expected to do in their positions. They should be compensated for managing, and their performance should be appraised in terms of management standards. Common sense as this is, very few companies, outside of the well-managed few, do it consistently.

The greater the divergence of the individual objectives of the members of the group from those of the group as a whole, the greater the leadership effort required to cause the group to act as a cohesive unit. In a new, young, or immature group, people will tend to do things their own way, to operate without consideration of the needs of others, and to make unilateral decisions. At this stage, the group does not know how to act as a team even though it may have outstanding individuals as members. Teamwork is developed most efficiently if the manager knows how to apply leadership effort so people will learn how to work together.

Institutional arrangements in the context of structure and strategy refer to the policies, systems and processes that organizations use to legislate, plan and manage their activities efficiently and to effectively coordinate with others in order to fulfil their mandate. What should be taken into consideration is an evaluation of both an appropriate institutional arrangement for the deep ocean, as well as an appropriate institutional arrangement for offshore and near shore maritime and coastal recreational activities. In searching for the most promising institutional arrangements it is perhaps worthwhile comparing the definitions of implementing bodies (see Table 1.1).

Table 1.1 Definitions of organizational and support institutions

Institution	*Definitions*	*Comment*
Alliance	• A union or association formed for mutual benefit and interest, especially between countries or organizations. • A relationship based on affinity of interests, nature or qualities.	Alliances are appropriate for the governance of ocean space shared by many countries but not dominated by any single country.
Assembly	• A group of people gathered together in one place for a common purpose. • The action of gathering together as a group for a common purpose.	An assembly is an appropriate structure for religious groups who gather in a church building, or for a scientific group who gather at a university, or a parliamentary group such as the National Assembly who gather to discuss politics.
Association	• A group of people organized for a joint purpose. • A connection or cooperative link between people or organizations.	Associations are generally appropriate for recreational clubs based on membership and chaired by an elected committee for a limited period of time.
Cluster	• A geographic concentration of interconnected companies and institutions working in a common industry.	Clusters are driven primarily by passion and commitment of the cluster champion and the participating stakeholders who collectively understand the benefits of collaborative economic development.
Collaborator	• A person who works jointly on an activity or project; an associate.	Collaborators are specialists who offer their professional services to different project teams and organizations mainly for short-term contracts.

Institution	*Definitions*	*Comment*
Council	• An advisory, deliberative or legislative body of people formally constituted and meeting regularly. • A body of people elected to manage the affairs of a city, country or other municipal districts.	A council is a formal organizational structure usually founded for a specific purpose with clear goals and objectives. Councils are usually comprised of respected leaders or people who have sufficient knowledge and experience to contribute to the effective governance of the council, e.g. a wise council or brains trust.
Organization	• An organized body of people with a particular purpose, especially a business, society, association, etc.	An appropriate legal entity for a structure governed by an elected management board or steering committee, with a constitution designed for either a private company or non-profit institution.
Partnership	• An association of two or more people as partners. • A business or firm owned and run by two or more partners.	A mutually beneficial agreement between two parties, either individuals or groups, who contribute their complementary but specific skills to strengthen a joint venture business association.

Lessons from tourism

Are there lessons to be learnt from tourism development that could be applied to the ocean economy? Tourism has been described as a complex, multifaceted and people-centred industry. It has also been likened to a fire and a double-edged sword. A fire can make you warm and cook your food but it can also burn your house down. Nowadays, tourism has become one of the biggest industries in the world. It provides jobs for 225 million people globally. Everything has its positive and negative sides but by analyzing the examples of real-life situations it is possible to evaluate the advantages and disadvantages of tourism.

To begin with there are benefits of tourism for both the tourist and the host destination. On a large scale it offers a good alternative to some more destructive industries for generating income both nationally and privately. The tourism industry encompasses many different areas, so it also creates jobs in many different areas. With tourism comes tour companies, hotels, car rental agencies, service stations, souvenir shops, different equipment rentals (sport equipment for example), and much more. All of this creates many different levels of employment for people in a given community. In many places the introduction and development of tourism allows local people an opportunity for economic growth that would not otherwise be available. In addition, it allows the tourist a chance to experience other cultures, which broadens understanding. Tourism is a great way for people to learn more and improve knowledge about the culture of other countries and nations.

On the other hand the tourism industry has a number of disadvantages. Tourism brings with it social problems. Tourism in certain areas can destroy the culture of the tourist location; old traditions may be lost or certain skills may be forgotten due to modernization, for example, residents forgetting how to farm. Noise from traffic can disturb local people, destroying their peace, calm and quiet.

Economic problems also come with tourism. For example in some popular locations, tourism raises the prices to a level unaffordable by local people. The exclusivity of the location can raise the prices of standard day-to-day products, as well as prices on real estate. Tourists require transport which increases the traffic flow and this can slow down normal traffic in some areas. Increased traffic can affect the delivery of products. In some places seasonal workers are required and these often come from outside of the tourism area. This means that the money earned by these contract workers is taken out of the local economy, resulting in money lost to local communities.

The disadvantage of tourism is that it can ruin the environment. Tourists are increasing the amount of pollution in many places. For example in recent years, traffic in the Himalayan region has increased drastically, mainly due to the desire of an increasing number of climbers wanting to ascend Mount Everest. Climbers leave their garbage everywhere, at different altitudes, including climbing equipment, foods, plastics, etc. In recent times, responsible tourism, green tourism and sustainable tourism are encouraged as best practice in many countries.

All things considered it can be said that tourism is indeed a double-edged sword. Tourism, like so many things in life, has its advantages and disadvantages. Humans need to learn to treat people and places with understanding and respect, without harming the environment.

Managing the tourism environment

Despite the negative side of tourism growth and development there are many excellent examples and success stories of how endangered wildlife species and vegetation types have been saved from extinction. Biodiversity planning, protected area management, tourism master plans and spatial development frameworks have allowed appropriate tourism development to progress within zones and nodes that determine what kind of tourism is appropriate according to sensitive ecosystems. Exactly the same principles and practices on terrestrial earth can be implemented in the oceans and seas. However, it is obviously much more difficult to implement responsible ocean governance because fences and protective barriers cannot be erected on the surface and most of the damage to the marine environment is below the surface.

Tourism structures and strategies

There is a lot to be learnt from tourism structures and strategies. In many developing countries, tourism structures have collapsed trying to emulate institutional arrangements in developed countries. The collapse is usually the result of a lack of a sustainable funding mechanism for regional and local tourism organizations combined with poor management performance of local and district municipalities. Bankrupt and financially stressed local municipalities divert funds dedicated for tourism development and promotion into other activities designed to keep voters happy. The ratio of public/private contributions to a local tourism organization in developing countries is on average about 90% public funding from the local municipality and 10% private sector funding from membership subscriptions. In contrast, about 60% of funding is contributed from a transient room tax or bed levy paid by overnight tourists in developed countries. The comparatively high failure rate of tourism structures in developing countries can usually be attributed to those structures being established without first considering the tourism strategy. Furthermore, the composition of board members is often not based on equitable public and private sector representation. A critical success factor is the quality of board members. At least 50% of the board members should have tourism industry experience, especially marketing, administration, operational and financial skills.

In many countries government-funded tourism structures have collapsed, which has forced the private sector to become more involved to the extent that private sector tourism associations have evolved and they are working independently of government. However, this has meant the demise of public–private tourism partnerships. By comparison, organized business is represented through respective chambers of commerce and industry both at the national and provincial levels. Tourism is a system, and for the system to function effectively it requires the public and private sectors to work together by forging mutually beneficial cooperative alliances. The ideal partnership should be built on a foundation of cooperation and coordination.

The New Zealand Tourism Board (NZTB) provides a joint venture fund to encourage the private sector to participate in joint venture programs. The NZTB tourism strategy is based on a key marketing principle: they need to give people compelling reasons to visit New Zealand. This means more of NZTB's spend must be directed to persuasive destination marketing. The NZTB accepts that they need to encourage greater cooperation and joint effort in the industry. Many tourism businesses individually do not have the budget or the resources to market themselves internationally.

By joining forces and clustering with other companies who appeal to the same type of visitor, some powerful arguments for visiting New Zealand can be presented. In this way, developing countries that have introduced regional tourism clusters driven by the private sector stakeholders but receiving enabling support from the national government ministries appear to be much more successful. Effective clusters are driven by passion and commitment. If stakeholders wish to withdraw from the cluster at any time, they can do so without any costs or legal implications.

Destination marketing

Although there does not appear to be a universal definition for destination marketing, the following description is relatively appropriate. People travel in relationship to a destination; destinations therefore influence their direction of travel as well as those activities engaged in once they arrive. Whether it is for leisure travel or planning a meeting, the destination becomes the critical planning variable. Understanding the important role that tourism marketing organizations play in formulating images and influencing travellers is fundamental for the smooth and efficient performance of the travel and tourism industry. The critical success factor is that the public and private sectors in the travel and tourism industry must forge a mutually beneficial cooperative alliance. These are the characteristics of a cluster, which are illustrated in the numerous case studies throughout this book.

Spatial development planning

Towards the end of the twentieth century it became obvious that sustainable tourism development must be integrated with destination marketing as there was sufficient evidence to prove that the mass tourism of the 1950s, which had very little regard for the environment, was being replaced by a new kind of tourism that was much more environmentally conscious.

Tourism is therefore expected to deliver economic benefits but not at the expense of sacrificing the environment. In fact tourism is now included in most Integrated Development Plans (IDPs) as a lead sector contributing significantly to local economic development (LED). In the early 1990s the concept of Spatial Development Initiatives (SDIs) and Collaborative Economic Development (Cluster Process) was introduced in many countries. Triple bottom-line measurement to monitor the social, economic and environmental impacts of, for example, global warming and climate change is now an imperative. Biodiversity management is another imperative aimed at managing land usage, effectively taking into consideration endangered species of flora and fauna. District and local municipalities are expected to prepare Spatial Development Frameworks (SDFs) to guide appropriate infrastructure development and to encourage inward investment.

All of the above have had a major impact on tourism planning. The World Tourism Organization (UNWTO) introduced the concept of Tourism Sector Plans (TSPs) in the new millennium when it was accepted that tourism was the best economic tool for poverty reduction in rural areas where development is needed the most. A TSP is an integrated tourism development and marketing plan in harmony with the environment. The goal of a TSP is to develop a strategy and action plan that will stimulate economic growth while protecting the environmental and heritage resources of a district or local municipality area thereby ensuring benefits to local people.

There are six main components of a TSP:

- Issues Analysis
- Situation Analysis
- Tourism Policy
- Tourism Spatial Development Plan
- Tourism Growth and Development Strategy
- Action Plan.

These components apply equally to responsible ocean governance, i.e. managing ocean activities within a marine spatial planning framework (maritime domain).

Regional cluster strategies

What is a cluster?

Formal definitions may vary, but many experts agree with economic expert and Harvard professor, Michael Porter's definition that a cluster is defined as a geographic concentration of interconnected companies and institutions working in a common industry (Porter 1998, 2000). In addition, clusters encompass an array of collaborating and competing services and providers that create a specialized infrastructure, which supports the cluster's industry. Finally, clusters draw upon a shared talent pool of specialized skilled labour. See Figure 1.1 for a typical maritime cluster. It is important to recognize however, that industry clusters are more than a group of firms within the same industry. The economic cluster model represents a synergy; a dynamic relationship and a network between not only the companies that comprise a cluster but also the successful partnering of the stakeholders. Government, education and other supporting organizations vital to a region's economic success represent these stakeholders. Many successful clusters have established a greater competitive advantage and wealth creation for their regions when compared to companies not in a cluster. Given this success, more policymakers and regions are considering fostering cluster development as building blocks of regional economies.

How do clusters form?

An industry cluster can start in a number of ways. Often a company's access to a labour force with specialized skills, a desirable location, plentiful natural resources, a supportive business infrastructure, or academic research may cause a group of companies to establish or cluster themselves and grow in a region. In other cases, face-to-face communication of business leaders can lead to cluster development and the creation of conditions that can aid in this process. Accelerating cluster growth and development is possible if business leaders explore the barriers and opportunities in a region that can foster cluster growth. Utilizing this qualitative approach helps to clarify the relationships within the industry and pinpoint

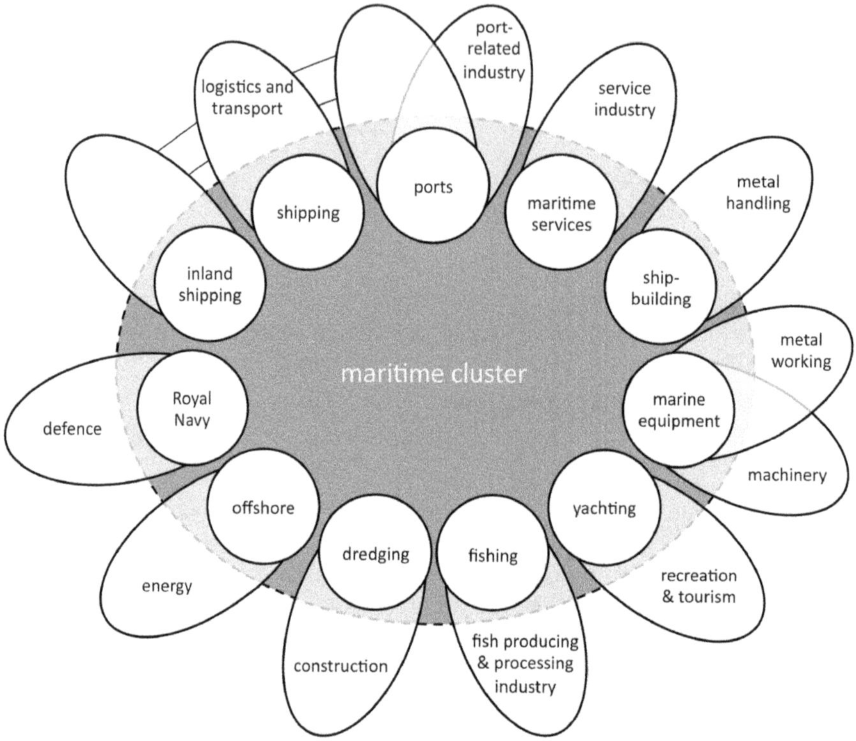

Figure 1.1 Typical maritime cluster

(*Source:* Dutch Maritime Network in Wijnolst (2006), with permission.)

specific infrastructure needs. Exploration in successful regions is industry led, has a high level of interactivity and has an ultimate goal of creating new collaborative partnerships between business, government, education and the community to support the demand infrastructure needs of the industry.

Creating a cluster framework

While it may take time to create thriving industry clusters, regions can take steps toward ensuring that an infrastructure is established that provides a foundation for cluster development. It is recommended that the following steps in creating a cluster framework should be demand driven, inclusive, collaborative, strategic, and value creating:

- Mobilize highly committed regional stakeholders and employers.
- Assess existing regional industry clusters.
- Foster collaboration by bringing together participants from key industries and institutions in the region.
- Implement the actions identified as a result of the collaborative process.

Benefits of clusters

Each region has its own cluster strength. A region that is able to successfully identify and work with its clusters will reap many strategic benefits. When examined from a global perspective, Wyman (2001) explains that regions offer significant competitive advantages:

> Regional clusters have the ability to offer local things such as knowledge, relationship, and motivation which cannot be matched by distant rivals.
> (Wyman 2001)

Therefore, regions that foster their local cluster-based strengths can reap tangible returns. Job creation and new entrepreneurial innovations are able to flourish in such a collaborative environment.

New leaders can develop within this context and continue to drive the networks and alliances of industry that will serve as power brokers within government and the community to continue to evolve a demand-driven framework. By working with clusters, community organizations and other support services may be able to increase their efficiency and effectiveness by directing services toward larger groups of firms. For example, organizations can enhance an industry's employee retention efforts through their childcare services, transportation services, housing and home ownership programs, and training programs. By working closely with a cluster, the organizations also have the opportunity to build relationships with industry that can facilitate the delivery of their services. Community organizations can work together with industry and public agencies to assist people in moving from welfare-to-work or school-to-work into specific industries with promising futures. By listening to industry cluster needs, economic developers can develop broad strategies to complement existing industries in the area. As a result, regional focus on such local strengths and industries can effectively and efficiently streamline marketing efforts, workforce development and produce greater return on regional investments.

Good governance

Ocean governance and the blue economy have become the new 'buzz words' in the ocean economy and maritime sector. It is therefore imperative that this book includes some thoughts on what constitutes good governance because certainly the best ocean strategies, policies and plans will have no impact unless implemented by responsible citizens of the blue planet. This implementation is complex because territorial boundaries are not visible, there are many maritime domains, decisions have to be made jointly, and laws are difficult to enforce. Hence responsible ocean governance has a lot to do with voluntary collaboration. In the last 20 years, the concepts of 'governance' and 'good governance' have become widely used in both the academic and donor communities. These two traditions have dissimilar conceptualizations. First, there is the academic approach, which focuses mainly on the study of the different ways in which power and authority relations are structured in a given society. Second, there is the donor community's approach, which puts emphasis on the role that state structures play in ensuring social, economic and policy equity and accountability through open-policy processes.

According to the academic approach, the generic understanding of governance is the management of resources and policymaking by means of exercising authority (power). Thus, it entails all instruments through which different policy stakeholders exercise legal rights with the aim to achieve political, economic, cultural and social objectives. In this sense, the term 'governance' appears to be more and more used in order to denote a complex set of structures and processes (at the public as well as at the private level), which are generally associated with national administration. However, its definitions offer a rather broad horizon of interpretation: wherever we can find this term, its definition varies slightly. For instance, in the report of the Commission on Global Governance (1995), governance is defined as:

> The sum of the many ways individuals and institutions, public and private, manage their common affairs. It is a continuing process through which conflicting or diverse interests may be accommodated and cooperative action may be taken. It includes formal institutions and regimes empowered to enforce compliance, as well as informal arrangements that people and institutions either have agreed to or perceive to be in their interest.

'Good governance' is a normative conception of the values according to which the act of governance is realized, and the method by which groups of social actors interact in a certain social context. The lack of a generally accepted definition of the concept is compensated by the identification of principles that strengthen good governance in any society. The most often enlisted principles include: participation, rule of law, transparency of decision making or openness, accountability, predictability or coherence, and effectiveness. The international donor community generally shares the view that these principles stand at the foundation of sustainable development. The first characteristic refers to equal participation by all members of society as the key element of good governance, with everyone having a role in the process of decision making. Secondly, good governance implies the rule of law is maintained through the impartiality and effectiveness of the legal system. Rule of law also means the protection of human rights (particularly those of minorities), independent judiciary and impartial and incorruptible law enforcement agencies.

The rule of law involves a variety of conditions, being strongly connected to good administration of justice, good legal framework, verified dispute mechanisms, equal access to justice, and the independence of judiciary workers (i.e. lawyers and judges). Good governance is also based on the transparency of the decision-making process, which ensures that information is freely available and accessible to those involved or affected by the decisions taken. Transparency therefore means free access to information. Last but not least, accountability and responsibility (of the institutions, just as much as of the civil society) are key requirements of good governance, with all of the participants in the political and economic processes being accountable for their decisions to each other.

The concept of good governance emerged mainly because practices of bad governance, characterized by corruption, unaccountable governments and lack of respect for human rights, had become increasingly dangerous, and the need to intervene in such cases had become urgent. Good governance has become an important element of the political and economic agendas, and has meanwhile been better specified along with the proliferation of good practices that take the concept into consideration. Beyond entailing just multiparty elections, a judiciary and a parliament, which have been identified as primary components of Western-style democracies, good governance goes on to address several other central issues:

> universal protection of human rights; non-discriminatory laws; efficient, impartial and rapid judicial processes; transparent public agencies; accountability for decisions by public officials; devolution of resources and decision making to local levels from the capital; and meaningful participation by citizens in debating public policies and choices.
>
> (Debiel and Terlinden n.d.)

A question of critical importance is how far can the concept actually go? The answer to this question can be found in practice rather than just in theory. Although the focus is on the public sector, where capacity building is called upon in order to strengthen and develop the social and economic situation, the concept also applies to the private sector and the civil society. When talking about good governance, what is generally referred to is the ability to perform efficiently, effectively and responsively, guided by principles that are feasible and desirable at all levels of the society, not just at the political one. As noticed in research material, reports and studies on the subject, good governance is a notion that refers to, and is functional in, a number of domains, going from institutional development to non-state actors' involvement.

Nowhere is this more evidenced than in the maritime zones established by the United Nations Convention on the Law of the Sea (UNCLOS), as agreed by some 157 countries at the time of writing. UNCLOS established a number of key definitions for delineating areas of the ocean:

- *Territorial Sea:* Water column and seabed extending 12 nautical miles from the coast.
- *Contiguous Zone:* Water column and seabed extending up to 12 nautical miles from the boundary of the territorial sea.
- *Exclusive Economic Zone (EEZ):* Water column and seabed extending up to 200 nautical miles from the coast.
- *Continental Shelf:* Relatively shallow area of seafloor adjacent to the coast that, in some instances, may extend beyond the EEZ.

- *High Seas:* Water column beyond the EEZ.
- *The Area:* Seabed beyond the EEZ.

According to Debiel and Terlinden (n.d.), the principle of good governance covers different ideals for a better society. Firstly, the *technical dimension* of the concept, which refers to 'the economic aspect of governance, namely the transparency of government accounts, the effectiveness of public resources management, and the stability of the regulatory environment for private sector activity'. Secondly, the *social dimension*, which is 'to build, strengthen and promote democratic institutions as well as tolerance throughout society'. And thirdly, the *political dimension* which refers to 'the legitimacy of government, the accountability of the political elements of government and respect for human rights and the rule of law'. When it comes to participation, equity, inclusiveness and transparency, minorities need a specific approach with regard to good governance strategies in terms of their relationship to the major stakeholders, since their position in the society is most often fragile.

When talking of minority issues and their involvement in the process of good governance, another subject comes to mind and is best exemplified in the Balkans, where the application of good governance principles in a post-conflict and divided society has been continuously on the political agenda since the 1990s. The principles of good governance are effective in post-conflict and divided societies as a neutral and practical set of measures meant to advance the rebuilding of the nation and to reset a vital connection between the civil society and the state institutions.

The reason why the basic principles of good governance are called upon in post-conflict societies is that it supposes the re-establishment of security and prosperity for all people, while managing the development of the state during the transition period. The main challenges refer to the process of constructing a democracy while surpassing the period of chaotic and precarious transition. The reason why the Balkans are an even more difficult and challenging subject for the supporters of the good governance principles is that the societies involved in this process need to cope not just with the institutional problems (e.g. lack of legitimate authority, a gap in the political tradition, etc.), but also with the social problems resulting from the internal conflicts between different ethnicities.

Thus, the principles of good governance practices need to ensure the development of *cooperative relationships* not only among different institutions, but also among different social groups. This communication among different actors and the participation of all in the governance process is an important prerequisite to cultivate the values that foster the respect for ethnic diversity, and support multicultural communication and the collaboration between representatives of different ethnic groups. Establishing a democratic system that builds on the structures of good governance principles is not an easy task, and has to take into account a number of complex issues that make every situation unique. Considerations include: the way the conflict was settled, the extent to which statehood had been destroyed or altered, and most importantly, the quantity of international support that contributes to the development of these countries. All of this has to be taken into account when strategies are proposed, in order to succeed in re-establishing peace and strengthening local and national capacities, together with the specificities of the area (mainly multi-ethnic cultures). Any strategy also needs to be managed through cooperation and respect for good governance principles.

After establishing the general configuration for rebuilding a post-conflict society on good governance architecture, the concrete measures can be operationalized by placing them into three different 'dimensions of governance'. These are identified by Debiel and Terlinden (n.d.) as: security, politics/administration and economy. The first and probably the most important prerequisite of good governance is security governance, which assures a framework for the application of all other conditions. Security problems appear in all post-conflict societies, as do problems such as lack of authority, inability of the legal institutions to control crime, erosion of the judiciary system, and use of force (by both the legally established and the underground organizations). In this case, civilian activities need to be strengthened and a division of authority has to be operated, in order to ensure transparency through public control and to create a secure environment for the people.

The second dimension, political–administrative governance, mainly refers to regaining the legitimacy of the government and of its decisions over the entire population. (This can be a major challenge, especially in post-civil war societies or multi-ethnic societies, where inclusiveness and transparency are two principles of upmost importance). In order to succeed in establishing a legitimate and transparent state, all groups have to feel represented, which can be accomplished if several guidelines are applied – either a 'horizontal division of power' or 'group autonomy and vertical division of power and forms of joint exercise of power through decentralization' (Debiel and Terlinden n.d.).

Thirdly, in order to assure a pragmatic framework, the socio-economic aspect also has to be considered, especially with regard to state reforms and the effectiveness of the newly established statehood. Stability and equity are two of the principles that should regulate all economic activity, with international cooperation being of particular special need, in order to adapt a former 'violence' economy to the structures of a peace-time economy.

CASE STUDY: WEST SWEDEN

The five clusters cross-sector collaboration model

The five clusters cross-sector model requires cross-boundary collaboration. Different regions can contribute in different ways depending on their specific strengths and resources. Thus, Gothenburg and West Sweden need to identify their strengths in order to determine how they can best contribute to growth and welfare, not only regionally and nationally but also globally.

Five strong clusters in West Sweden have been identified jointly by leading representatives from the academia and the public and private sectors. The clusters each have a long tradition, and they are all facing some exciting challenges. The five clusters are:

- Urban future
- The marine environment and the maritime sector
- Transport solutions
- Green chemistry and bio-based products
- Life science.

The intention is for these clusters to develop cross-boundary collaborations, something Gothenburg and West Sweden have always done. The region is characterized by openness to the surrounding world, both nationally and internationally. It is also known for the closeness between the academia and the public and private sectors, not to mention the openness among people in general. However, in order to be successful they have to try new approaches. In addition to their renowned cooperative spirit, they must reinforce the sense of trust that effective cooperation is built upon.

Mistra Urban Futures

West Sweden is a forerunner in cross-sector collaborations concerning sustainable urban development. Several established processes and arenas for cooperation, capacity building and knowledge creation are in place. West Sweden is also a seedbed and a base for a large number of innovative companies that in one way or another are engaged in sustainable urban development. The region offers these companies ample opportunities to develop and test tomorrow's products and services in collaboration with other actors.

Mistra Urban Futures is a new center for sustainable urban development with an ambition to become a world leader in the field within a few years. The center, which is managed from Gothenburg, was formed to find out more about the difficulties and opportunities the world's cities are facing. It consists of five local platforms for cooperation located around the world: Gothenburg (Sweden), Cape Town (South Africa), Kisumu (Kenya), Manchester (UK) and Shanghai (China). Mistra believes that close cooperation between practitioners and researchers will yield first-class knowledge. The center will contribute to making a real difference for the environment and for people's lives in the cities of the world. The primary strength of Mistra Urban Futures is the close cooperation between research and practical application. The center offers an arena for development and transmission of knowledge, where cooperation with the private sector, interest groups and the public will be developed. It is a forum for knowledge and will develop at pace with the flow of ideas and experiences that the involved actors contribute.

River City Gothenburg

River City Gothenburg refers to the areas in central Gothenburg (Centrala Älvstaden) that are located on both sides of the river that flows through the city. A unique feature of the area is that there are empty lots in the city center that can be developed. One of the challenges will be to tie the city together across the river. There is broad agreement regarding how Gothenburg and the Gothenburg region should be developed. The central parts of Gothenburg, the regional core, should be invigorated in order to strengthen West Sweden in general. A vision and a strategy for the large central areas that are going to make up the regional core in the future are currently being developed. The areas – former harbor, terminal and industrial areas on both sides of the Göta Älv river – all have good potential to be turned into good examples of sustainable urban development and serve as a testing ground for new innovative ideas.

Academia and the public and private sectors are collaborating to build River City Gothenburg. The purpose is to ensure that River City Gothenburg is given high-quality sustainable urban development by developing work models for collaboration, dialogue and information. The intention is to collect experiences gained elsewhere, test them on River City

Gothenburg, and then document the experiences so that they can be shared both nationally and internationally. It is a matter of creating urban development that is inclusive, that contributes to a complex and robust economy, and that can handle the effects of climate change. Children are the future, and all urban development in Gothenburg should be based on their needs. The City of Gothenburg has, in broad collaboration with other actors, successfully developed models aimed to promote the social sustainability dimension. The tools that have been developed for child and social impact assessments are implemented in all development efforts related to the city's physical environment.

The marine environment and the maritime sector

The sea is one of West Sweden's core assets and entrepreneurship linked to the marine environment has a long history. Herring fishing, shipbuilding, ocean trade and port operations are some examples. The maritime cluster as a whole is well rooted in both academia and the public and private sectors of the region. However, despite the remarkably large number of local companies that have proven to be internationally competitive in the maritime sector, there is generally a shortage of maritime knowledge. Thus, there is an urgent need for research and the creation of knowledge in order to take full advantage of the diversity of the marine environment. For this reason, marine research has a strong position at both the University of Gothenburg and Chalmers University of Technology.

There are numerous global challenges in the maritime sector. In a nutshell, it is a matter of the essential conditions for life. For example, the sea must be able to remain the most important global source of protein, the sea needs to be used as a renewable source of energy, and sustainable transportation must be used in the sea without harming the environment. The University of Gothenburg and Chalmers University of Technology are involved in all of these areas, as well as in more basic research aimed to facilitate sustainable use of marine resources, such as research on the genetics of marine organisms.

Since these global challenges need to be solved jointly by all countries in the world, the maritime cluster in West Sweden is characterized by both partnerships with competence centres around the Baltic and North Seas and by collaborations on a larger European or global scale. The maritime strategy for West Sweden emphasizes the importance of cooperation for the creation of innovative environments. Academia and the public and private sectors collaborate in several strategic maritime projects aimed both to strengthen the competitiveness of the maritime sector and to meet society's challenges in a global perspective.

The ample access to broad competences that include the perspectives of academia and the public and private sectors on local, regional and state levels implies good prospects for cooperation in the maritime sector. One interesting example concerns fishing in the northern parts of Bohuslän, where fishermen and researchers, with the support of authorities, are cooperating towards a shared vision to make the fishing interfere less with the unique marine environment in the Koster Sea, which is full of corals and other species that cannot be found anywhere else in the country. In the cooperation, fishermen are studying marine biology, and researchers are joining fishing crews in their work in order to learn more about it. As a result, shrimp fishing is now harmless to the marine environment.

When the marine national park Kosterhavet opened in 2009, the coastal fishing could continue thanks to the creative collaboration between the fishing community and researchers. The coastal environment with its productive shallow areas is particularly sensitive to emissions of, for example, sewage and fertilizers from farming. Researchers and entrepreneurs within land and sea management were pioneers within mussel farming, working to reduce the effects of over-fertilization and to produce high-quality food products. Shellfish and mussels from the coast of Bohuslän are exported in large quantities to continental Europe.

Although the marine aquaculture along the west coast of Sweden is largely based on mussel farming, it also has a potential for ecological fish farming. Marine research, high environmental standards in food production and access to clean water imply unique conditions for experimental marine aquaculture. Researchers at the Centre for Aquaculture at the University of Gothenburg, together with representatives of the private sector, are exploring the potential for fish farming in the marine environment. Modern aquaculture is characterized by sustainable utilization of biological resources from the sea. Norwegian and Swedish researchers are collaborating on the sustainable use of bio-marine resources in a joint project on Skagerrak's unknown flora and fauna. The aim of the study is to utilize features of marine organisms that may be beneficial in the production of materials, medicines and nutritious food additives.

The sustainable use of marine resources requires new equipment and tools. EU requirements communicated through various directives, and the introduction of marine spatial planning, are pushing forward the development of marine measurement technology. In this context it should be noted that West Sweden has the possibility to offer a testing ground for international companies due to its marine field stations and technical expertise.

Another example of how effective collaboration can yield creative solutions to complex problems concerns fouling, which is the growth of microorganisms, plants, algae and even animals on the hull of a boat or ship. The heavy metals and chemicals that have been used in the past to prevent fouling have led to emissions of toxins into the sea and severe damage to the fauna. Joint efforts by the University of Gothenburg and Chalmers University of Technology have resulted in anti-fouling methods that do not harm the environment. The Gothenburg-based company I-Tech has developed an anti-fouling biocide, Selektope, which does not interfere with the surrounding environment. As a result of cooperation with Volvo Penta, the new anti-fouling substance is ready to enter the global market. I-Tech's innovation may make the company a key provider in the anti-fouling market, which totals some 2.5 billion SEK (£2.2 billion) per year.

Benefiting from a windy coastal location it is an explicit goal of West Sweden to become fossil independent by 2030. This will require a transformation of large parts of the energy supply. Fossil fuels must be replaced by renewable energy on a large scale. The sea is an enormous source of energy, and researchers are working intensively to resolve how best to control and use it. There are plans to generate 10% of Europe's total power supply by means of offshore wind power by the beginning of the 2020s. General Electric (GE), a world-leading hi-tech company, placed an offshore wind power prototype in Gothenburg in the winter of 2011/12. The project is a collaboration involving GE, Göteborg Energi, Chalmers and the Swedish Energy Agency. However, the technical and financial challenges linked to offshore energy extraction are considerable, and there are no plans to put this model into serial production.

Large-scale development of wind- and wave-power parks requires a number of test sites. There are significant synergies between wind and wave power. Connection to the main power grid, equipment maintenance in a demanding marine environment, and materials for and sustainability of marine energy extraction are some areas where knowledge in the wind power sector can be used at the Ocean Energy Centre in Gothenburg. In order for ports to be located near business, industry and large cities, the emissions from sea transportation must be reduced. One of the most important methods to accomplish this is to connect the ships to the main power grid.

The Port of Gothenburg is one of the world leaders for onshore power supply connection and was first in the world to provide high-voltage onshore power supply for cargo ships. Another step towards greener sea transport in Gothenburg was taken when Stena Line opened a new onshore power supply facility for its ferries operating between Sweden and Germany. In total, one in three vessels that enter the Port of Gothenburg can now turn off their diesel engines at the quayside and instead use renewable power from land-based wind power. The Port of Gothenburg's successful investments in transportation to and from the port via rail is another example of a sustainable logistics solution.

Conclusion

West Sweden is the natural Swedish base for the industry that deals with transportation and logistics to Sweden as a whole and to large parts of the rest of Northern Europe. The region is home to all the actors that in the future may help develop new sustainable and innovative solutions that require cooperation and a holistic approach. This makes it possible to test solutions that require cooperation, such as energy efficiency measures, new fuels, and security and logistics systems, before they are implemented on a larger scale. Research centers, transportation companies and subcontractors can come together, test and further develop new solutions at their shared testing ground Lighthouse, a cross-disciplinary competence center situated at Lindholmen Science Park. The center is based at Chalmers University of

Technology and the University of Gothenburg, but is strongly supported by the public and private sectors.

Note: Five clusters is a collaborative initiative between Chalmers University of Technology, City of Gothenburg, University of Gothenburg, Region Västra Götaland and West Sweden Chamber of Commerce.

Discussion

Managing the multiple uses of the ocean is vital. Successful management of any activity, however, requires adequate understanding both of the activity and of the context in which it takes place. Such an understanding is needed even more when the management tasks are divided among many players: unless each knows how the part they have to play fits into the overall pattern, there are risks of confusion, contradictory actions and failures to act. Managing human uses of the ocean has inevitably to be divided among many players. By the end of this book readers should be in a position to make an informed judgement with respect to appropriate institutional arrangements for responsible governance in the ocean economy.

References and further reading

Commission on Global Governance (1995). *Our Global Neighbourhood: The report of the Commission on Global Governance*. Oxford: Oxford University Press.

Debiel, T. and Terlinden, U. (n.d.). 'Good governance – The concept'. www.edrc.ro/docs/docs/11-17.pdf (accessed: May 2017).

European Commission (2016). *Study on Specific Challenges for a Sustainable Development of Coastal and Maritime Tourism in Europe*. Luxembourg: Publications Office of the European Union.

Gdynia Maritime University (2016). *Polish Maritime Cluster and the Polish Economy*. Gdynia: Gdynia Maritime University.

Maritime Cluster of West Sweden brochure (n.d.).

Porter, M.E. (1998). 'Clusters and the new economics of competitiveness'. *Harvard Business Review*, December: 77–90.

Porter, M.E. (2000). 'Location, competition, and economic development: Local clusters in a global economy'. *Economic Development Quarterly*, 14: 15–34.

Region Västra Götaland, University of Gothenburg, Chalmers (2012). 'Innovation and development of the maritime clusters in Västra Götaland: Project presentation'. www.gu.se/digitalAssets/1378/1378432_5042_maritime_cluster_folder_eng.pdf (accessed: April 2017).

Swinsky Roques Dias, D. (2013). 'The Porter Report revisited: Creating and assessing a cluster evaluation framework'. https://fenix.tecnico.ulisboa.pt/downloadFile/395146006381/Dissertation%20-%20Daniel%20Swirsky%20Roque%20Dias.pdf (accessed April 2017).

Wijnolst, N. (2006). *Dynamic European Maritime Clusters*. Delft: Delft University Press.

World Maritime News (2016). 'In depth: The Dutch maritime cluster monitor 2016'. http://worldmaritimenews.com/archives/211469/the-dutch-maritime-cluster-monitor-2016/ (accessed: April 2017).

Wyman, M. (2001). 'Thinking about governance: A draft discussion paper'. http://citeseerx.ist.psu.edu/viewdoc/download;jsessionid=85FD124C29BADFA2909166CD52B77C6B?doi=10.1.1.199.5627&rep=rep1&type=pdf (accessed: May 2017).

Chapter 2

The growing multiuse ocean

The ocean is an interconnected global ecosystem supporting a wide range of uses. Maintaining a healthy ocean requires responsible stewardship by all users. The ocean supports a growing range and level of commercial and other activities. User conflicts are on the rise, as are impacts from increasing use. However, the best efforts by a single company or an entire industry will not be enough to address cumulative impacts in the interconnected marine ecosystem. Companies dependent on the ocean will benefit from collaboration with others in the ocean business community to address marine environmental issues. The responsible governance of integrated maritime policies and strategies requires a high level of multi-stakeholder collaboration both onshore and in the deep ocean.

The EU's Integrated Maritime Policy

Europe's seas and oceans are a rich and often underestimated source of innovation, growth and employment. They provide valuable ecosystem services and resources on which all marine activities depend. Since its creation in 2007, the Integrated Maritime Policy (IMP) has sought to enhance the sustainable development of the European maritime economy and to better protect the marine environment by facilitating the cooperation of all maritime players across sectors and borders.

Five years later, the economic climate has radically changed. With the Europe 2020 strategy, the EU is seeking to get the European economy back on track to deliver employment, competitiveness and social cohesion. The potential for growth in the maritime economy is an opportunity that Europe, as a maritime continent, needs to seize. Since 2009, the EU has launched key initiatives in all policy areas related to the seas in order to strengthen Europe's competitiveness. By avoiding duplication of spending and efforts, and encouraging the sustainable development of maritime activities, the IMP has brought concrete benefits to the European economy and maritime sectors in member states.

Europe is going through cuts in public spending, so delivering maximum results with limited means is crucial. Cooperation makes operations at sea more cost efficient and optimizes data usage. As new uses of the sea are being developed daily, it is crucial that member states put in place stable planning systems favouring long-term investment and cross-border coherence. Investment in research must deliver fully its potential for innovation in the maritime economy. The EU strategy for marine and maritime research has helped achieve this objective. With the Marine Strategy Framework Directive, our seas and oceans can be managed sustainably. Sustainability is a pre-condition to developing the maritime economy and is driving advanced technologies that strengthen Europe's long-term competitiveness.

In 2009, the Council and European Parliament welcomed the first Progress Report on the IMP and asked the Commission to present further developments in 2012. This second Report describes the progress of the EU's IMP and maritime sectoral policies between 2010 and 2012. It highlights their contribution to the objectives of the Europe 2020 strategy by taking stock of concrete achievements that help create growth, enhance resource efficiency and save public money.

The contribution of the maritime economy to growth and employment

Blue growth

In September 2012, the Commission adopted a Communication on Blue Growth; opportunities for marine and maritime sustainable growth. The objective was to launch a joint initiative with member states, regions and all relevant stakeholders to unlock the potential of the blue economy.

A study launched by the Commission in 2010 identified as common bottlenecks the lack of appropriate skills, access to risk capital, fragmented marine data, environmental challenges and difficult planning processes. Actions are underway in member states to remedy the skills shortage, for example through encouraging clusters of industry and education. Integrated maritime policy initiatives are addressing the data and planning issues. However, certain sectors require a more targeted approach. Accordingly, five areas with growth potential – maritime and coastal tourism, ocean renewable energy, marine mineral resources, aquaculture and blue biotechnology – have been analyzed in more depth to suggest additional measures needed to increase growth and jobs.

Maritime transport

Maritime transport services are essential for the European economy to compete globally. In 2011, the Commission adopted a White Paper for Transport. It further specified the orientations of the Maritime Transport Strategy until 2018: the ability to provide cost-efficient maritime transport services; the long-term competitiveness of the EU shipping sector; and the creation of seamless transport chains for passengers and cargo across transport modes. Following the 2009 Communication on a European maritime transport space without barriers, the Directive on reporting formalities entered into force in 2010. This initiative simplifies and harmonizes administrative procedures in order to boost intra-EU maritime transport. In 2011, the Commission proposed new guidelines for Trans-European Networks to broaden the role of the Motorways of the Sea as main European corridors. Through multiannual calls, the Commission is leading the way in reducing the environmental impact of transport and in increasing transport efficiency.

Energy

Europe's citizens, industry and economy depend on safe, secure, sustainable and affordable energy. Offshore wind energy contributes to a 20% share of energy from renewable sources by 2020. It is a priority of the EU's Strategic Energy Technology Plan, through which industry, member states and the Commission work on a long-term approach to technology development and demonstration. The Research Framework Program and the Intelligent Energy Program further support the development of wind and ocean energy technology, which contributes significantly to growth in coastal regions. In 2011, the Commission proposed guidelines to lay down rules for the development and interoperability of trans-European energy networks. Priority corridors were identified, including the North Sea Offshore Grid and the Baltic Energy Market Interconnection Plan.

Shipbuilding

European maritime industry with its strong innovation and design capacity has a strategic role to play in addressing challenges such as climate change, air pollution, energy efficiency and development of offshore activities. As a strategic response to the challenges in the sector, the initiative LeaderSHIP fosters competitiveness based on EU technological leadership segments. The initiative is currently under review to address future opportunities by greening of shipping and a diversification into new business areas such as offshore wind energy. The Framework on State Aid to Shipbuilding determines which types of state aid are allowed for shipyards. A new Framework was adopted in December 2011 and was valid until 2013. It contained specific provisions in relation to innovation aid and regional aid for shipbuilding, as well as provisions on exports credits. Its extended scope applied to inland waterway vessels and floating and moving offshore structures.

Fisheries and aquaculture

EU fisheries are affected by several interconnected problems: fish stocks are overfished; the economic situation of parts of the fleet is fragile, despite high levels of subsidies; jobs are unattractive; and the situation of many coastal communities depending on fisheries is precarious. In July 2011, the Commission adopted a package of initiatives, including new legislative proposals, to reform the Common Fisheries Policy. It aims to provide the building blocks for sustainable fisheries while respecting the ecosystem as well as ensuring quality food supplies, thriving coastal communities, profitable industries, and attractive and safer jobs. Long-term management with clear sustainability targets for the exploitation of resources and the stopping of wasteful practices are at the heart of the proposals. Support will also be given for improving data to underpin policy choices and to ensure better enforcement and control. The transition will be accompanied by a European Maritime and Fisheries Fund which aims to improve sustainability and the performance of small-scale coastal fisheries, to aid the promotion of aquaculture, to support job creation in maritime communities, and to deliver cost-efficiency in maritime affairs.

EU cooperation to ensure maritime economy growth

The IMP sets out a coherent strategy to enhance the sustainable development of maritime sectors. Coordination across sectors and borders is necessary to manage the increasing impact of maritime activities on each member state and on the environment, to ensure the safety and security of European citizens and to maintain a qualified workforce.

Maximizing the sustainable deployment of activities on coasts and at sea

Sectors compete for space and resources across sea basins. Conflicts between sea uses and demands for sea space will continue to increase, in particular due to emerging offshore activities. As Europe is going through a severe economic crisis, tools are needed that enable growth by facilitating the coexistence of multiple activities, while reducing environmental impacts. Maritime Spatial Planning (MSP) and Integrated Coastal Zone Management (ICZM) ensure the efficient cross-sectoral and cross-border planning of marine waters and management of costal zones. They are crucial to guaranteeing sustainability, providing legal predictability and reducing costs for investors and operators, in particular those operating in cross-border areas.

Advances in delivering national ICZM strategies have been made, but reports received from 16 EU member states in 2011 showed large variations. In 2011, the ICZM Protocol to the Barcelona Convention entered into force, making ICZM compulsory for coastal member states in the Mediterranean. The Commission acts as a facilitator for the development of a common framework for MSP within the EU, and delivered a legislative proposal on MSP and ICZM at the end of 2012. Studies on the economic benefits and impact of MSP and ICZM, two pilot projects on cross-border cooperation and the Ourcoast platform on best practices for ICZM have contributed to the proposal.

The territorial benefits of maritime policy

Coastal regions have key renewable resources for maritime growth, including hydro, wave, wind, tidal and biomass energy. Coastal tourism depends on attractive and healthy marine environments. As social exclusion can be particularly acute in coastal areas and islands, developing an inclusive maritime economy is essential. The need for integrated marine spatial planning is a global imperative. Ideally, terrestrial spatial planning should flow into marine spatial planning with coastal tourism management programs sandwiched in-between. This would ensure that land-based activities, and onshore, near-shore and offshore coastal and marine tourism activities, are in harmony with the terrestrial and marine environment.

Regional policy

Europe is both a territorial and maritime entity. For maritime sectors to support growth on land, connectivity must be increased. The 2011 Communication Regional policy contributing to smart growth in Europe 2020 called for national and regional governments to set up Smart Specialization Platforms on research, regional, enterprise, innovation and education policies. This approach can better coordinate programming aimed at the development of coastal regions and their economies. A further step is the proposal for Common Provisions and a Common Strategic Framework to link investment priorities under cohesion, rural development, fisheries and maritime policy between 2014 and 2020. Numerous maritime projects have been supported through the EU regional policy funds. Examples include desalination units or offshore infrastructure projects for the demonstration of wave energy generation. Several EU Cross-Border Cooperation Programs have also promoted maritime growth, including projects looking into algae as a potential source of biofuel and best practices in climate change adaptation in coastal regions.

Sea basin strategies

Transnational cooperation at sea basin level is an efficient approach to the development of the maritime economy and to the protection of the marine environment. Sea basin strategies cater for the economic characteristics of Europe's basins, while allowing better use of public money. By aligning existing European and national funds on agreed growth priorities, they act as economic drivers in maritime regions. The EU strategy for the Baltic Sea Region has yielded more than 80 flagship projects whose detailed impact is described in the Progress Report adopted in June 2011. The strategy includes measures to reduce pollution from vessels, to develop sustainable short-sea shipping, to tackle eutrophication, to create new clusters for innovative small and medium-sized enterprises (SMEs), to support research, and to integrate maritime surveillance systems.

A Maritime Strategy for the Atlantic was adopted in November 2011 to boost job creation and growth in the Atlantic area by enhancing its maritime potential. An Atlantic Forum is identifying priority actions through an Action Plan adopted in 2013. It allows the strategic use of EU structural funding to support maritime growth for the period 2014 to 2020.

A Communication for better maritime governance in the Mediterranean was adopted in 2009 to improve governance of maritime affairs and environmental protection. Regions and member states have redirected some 2007–2013 funding for European territorial cooperation to maritime objectives. Technical assistance to support growth from the sea and improve maritime policy making in non-EU partner countries is now being provided through the Integrated Maritime Policy in the Mediterranean (IMP-MED) project under the European Neighbourhood Policy.

Italy, Slovenia, Greece and Croatia are engaged in deeper maritime cooperation at sub-regional level in the Adriatic and Ionian Seas. A maritime strategy is under development to define priority areas for growth and to streamline EU funding accordingly. Cooperation around the Black Sea is likewise moving ahead, following a high-level brainstorming event organized with Bulgaria and Romania in October 2011.

In June 2012, the Commission and the EU High Representative for Foreign Affairs and Security Policy adopted a Communication proposing 28 action points for the EU's constructive engagement in the Arctic. It supports effective stewardship of the Arctic based on knowledge, responsibility and engagement to meet its increasing strategic, economic and environmental challenges.

Protecting marine ecosystems – A condition and factor for growth

The EU has the largest maritime territory in the world. Ensuring the health of marine ecosystems is necessary for the future of ocean biodiversity and to sustain maritime growth. As European maritime companies are frequently top performers in innovative environmental technology, environmental legislation also supports economic development.

The challenge of healthy marine ecosystems

The 2008 Marine Strategy Framework Directive is the environment pillar of the IMP. Its overarching aim is to achieve good environmental status (GES) for EU marine waters by 2020. A number of milestones have been completed in order to reach this aim. In 2010, the Commission adopted a Decision on criteria and methodological standards for the assessment of GES, which will be used by member states in developing monitoring programs and cost-effective measures to ensure the GES of their marine waters. In 2011, the Commission also clarified the relationship between the initial assessment of marine waters and the criteria for GES. Significant progress has been achieved in the establishment of the Natura 2000 network, but gaps still exist, especially offshore. In 2011 the Commission adopted Guidelines on the Implementation of the Birds and Habitats Directives in Estuaries and Coastal Zones with particular attention to port development and dredging. The Commission's LIFE+ financial instrument contributes to managing the Natura 2000 network in the marine environment by promoting innovative protection measures and capacity building.

Adaptation to and mitigation of climate change

Climate change can have dramatic consequences for coastal regions, including threats to coastal defences, erosion, flooding, and rising sea levels, and can have higher impacts in combination with other pressures on the marine environment. In March 2012, the Commission launched the European Climate Adaptation Platform, the most comprehensive website for information on climate change impacts and vulnerabilities in Europe. It aims to support policymakers in the development of climate change adaptation measures, including in coastal areas.

Addressing air pollution from ships

Greenhouse gas (GHG) emissions from international maritime transport currently represent around 3% of global GHG emissions. This is likely to grow due to increasing world trade and demand for shipping. In 2011, the Commission set out the goal of reducing the EU's GHG emissions from maritime transport by 40% by 2050. Cooperation projects to monitor GHG emissions from ships in Europe have been launched by the European Maritime Safety Agency, the European Environmental Agency, the Finnish Meteorological Institute and the International Maritime Organization. In 2011 the Commission also adopted a proposal for an amendment of Directive 1999/32/EC to significantly reduce shipping emissions of sulphur and to contribute to reduce air quality problems in the EU which affect human health and acidification.

Better management of maritime affairs

Good management and coordination within and among member states, coastal regions, industries and stakeholders is essential to develop the maritime economy as efficiently and sustainably as possible. A flow of information, data and best practices can accelerate investment and innovation while contributing to better environmental protection.

Developments in EU member states

Member states are increasingly applying coordinated approaches for the development of their maritime economy, through national strategies, such as in France, Portugal and Germany, or through specific initiatives, such as the UK Marine Bill, the Danish Maritime Strategy or the Irish Science Strategy. In 2011, Germany also adopted a plan to boost its maritime economy. A number of member states have set up inter-ministerial coordination or established ministerial posts, such as Cyprus, France, the Netherlands and Poland. Maritime regions such as Schleswig-Holstein, Västra Götaland and Brittany have further developed their regional maritime strategies since 2009. In 2010, 17 member states and associated countries launched the Joint Programming Initiative Healthy and Productive Seas and Oceans to promote synergies between research resources and capacities.

Developments at international level

In line with the 2009 Communication on the international dimension of the IMP, the Commission has strengthened its efforts in the international maritime arena. At global level, the EU has pushed for more ambition in the Resolutions on Oceans and the Law of the Sea and on Sustainable Fisheries, promoting global membership of maritime governance instruments such as The United Nations Convention on the Law of the Sea (UNCLOS). A particular success is the launching of a process at UN level which should ultimately lead to the negotiation of an UNCLOS implementing agreement for the conservation and sustainable use of marine biodiversity in areas beyond national jurisdiction. The EU also pushed for progress in the protection of oceans and seas, and in maritime governance at the UN Conference on Sustainable Development in Rio in June 2012. Bolder external action is yielding success, with stronger performance by Regional Fisheries Management Organizations and increasing cooperation with less developed countries in the fight against illegal, unreported and unregulated fishing. Maritime affairs have become a regular topic in discussions with the EU's partners, such as China, Russia, Japan, Canada and the US. The scope of sectoral dialogues has been gradually expanding to embrace more overarching cooperation on global maritime affairs.

Awareness and visibility of maritime Europe

A first aim of the IMP was to raise the visibility of Europe's maritime identity and economic potential among Europeans. Public awareness is essential in areas where maritime interests compete with other sectors for political support, investment or the labour market. Since the launch of the IMP, Eurostat has been involved in producing statistical information to support the policy so as to improve economic data on maritime sectors and maritime regions. Maritime initiatives have been included in Eurostat work programs. Since 2009, statistics have been regularly published for coastal regions and maritime sectors. Interactive tools like the Maritime Forum or the Atlas of the Sea facilitate access to sea-related information and help to increase knowledge of maritime Europe. The IMP was created to reaffirm the maritime dimension of the EU. It has grown from this vision into a tool delivering concrete benefits for maritime growth and sustainability in Europe. The strong contribution of maritime sectors to Europe's economy, and the objectives of the Europe 2020 strategy, is reinforced by coordinated action to lower costs, improve resource efficiency, reduce risks, support innovation and make better use of public money.

Impacts on biodiversity

In an unprecedented effort that was published online (Dalhousie University 2010), a team of scientists mapped and analyzed global biodiversity patterns for over 11,000 marine species ranging from tiny zooplankton to sharks and whales (Tittensor et al. 2010). The researchers found striking similarities among the distribution patterns, with temperature strongly linked to biodiversity for all 13 groups studied. These results imply that future changes in ocean temperature, such as those due to climate change, may greatly affect the distribution of life in the sea. The scientists also found a high overlap between areas of high human impact and hotspots of marine diversity. Much research has been conducted on diversity patterns on land, but our knowledge of the distribution of marine life has been more limited. This has changed through the decade-long efforts of the *Census of Marine Life* (www.coml.org), upon which the research builds. The authors synthesized global diversity patterns for major species groups including corals, fishes, whales, seals, sharks, mangroves, seagrasses, and zooplankton. In the process, the global diversity of all coastal fish species has been mapped for the first time.

The researchers were interested in whether there are consistent 'biodiversity hotspots' – i.e. areas of especially high numbers of species for many different types of marine organisms simultaneously. They found that the distribution of marine life showed two fundamental patterns: coastal species such as corals and coastal fishes tended to peak in diversity around Southeast Asia, whereas open-ocean creatures such as tunas and whales showed much broader hotspots across the mid-latitude oceans. They also tested whether these global patterns could be consistently explained by one or more environmental factors. Temperature was the only factor found to be linked with the distribution of all species groups, with the availability of habitat also playing a role.

The study observed how consistently temperature was linked with marine diversity. This relationship suggests that ocean warming, such as that due to climate change, may rearrange the distribution of oceanic life. The researchers noted that while we are increasingly aware of global gradients in diversity and their associated environmental factors, our knowledge of patterns in the ocean has lagged behind that of patterns on land. The study attempts to help overcome this disparity. It also assessed the overlap between hotspots of marine diversity and human impacts, i.e. the combined effects of fishing, habitat alteration, climate change and pollution. Human impacts were found to be particularly concentrated in areas of high diversity, suggesting the potential for severe species losses in these regions. The combined effects of exploitation, habitat alteration, pollution and climate change are threatening the diversity of life in the global ocean. Collaborative research studies provide further evidence that limiting ocean warming and other human impacts will be particularly important in securing these hotspots of marine biodiversity into the future.

Biodiversity and the functioning of ecosystems are often tightly coupled, with highly diverse ecosystems providing more goods and services that benefit human beings, as well as being more resilient in the face of disturbance, than less diverse ecosystems. The observed concentration of human impacts in our richest marine areas is a worrying indication of our growing footprint in the oceans. The author argues that effective collaboration between marine scientists, coastal authorities, ocean business communities, coastal and marine tourism practitioners, and academics is the only way to ensure a multidisciplinary approach to solving ocean-related issues. In this regard a maritime cluster is an excellent special purpose vehicle (SPV) for achieving well-informed, consultative and collective decisions.

CASE STUDY: ENGLAND, THE JURASSIC COAST

Introduction

This is England's only natural World Heritage Site (WHS). The 95 miles (155 km) of Dorset and East Devon coast was designated in 2001 by UNESCO. The primary reasons for designation were:

- The sequencing of 185 million years of geological history based upon the Jurassic, Triassic and Cretaceous periods.
- The contribution the area has made to earth sciences and understanding geology, especially as a result of the fossil beds.
- The remarkably varied coastal scenery and geomorphology, with unique features such as Chesil Beach.
- The combination of important habitats and conservation features.

Location

The WHS-designated area (see Figure 2.1) incorporates estuaries, islands and coastlines, and also includes a number of resorts, towns and villages, with a population of 440,000. Tourist resorts have played an important role in developing the area's economy for 200 years, including Bournemouth, Poole, Swanage, Weymouth, Lyme Regis and Exmouth. The coastline is also characterised by numerous small attractive villages and small ports such as West Bay, Seaton, Charmouth and Burton Bradstock.

In addition to the WHS designation, the area incorporates two Heritage Coasts, as well as 13 Sites of Special Scientific Interest, several National Nature Reserves, four harbours (including Poole, which is a Special Protection Area and a wetland of international importance – RAMSAR – site), a Special Area of Conservation (Chesil Beach and the Fleet), and part of the Dorset coast is an Area of Outstanding Natural History.

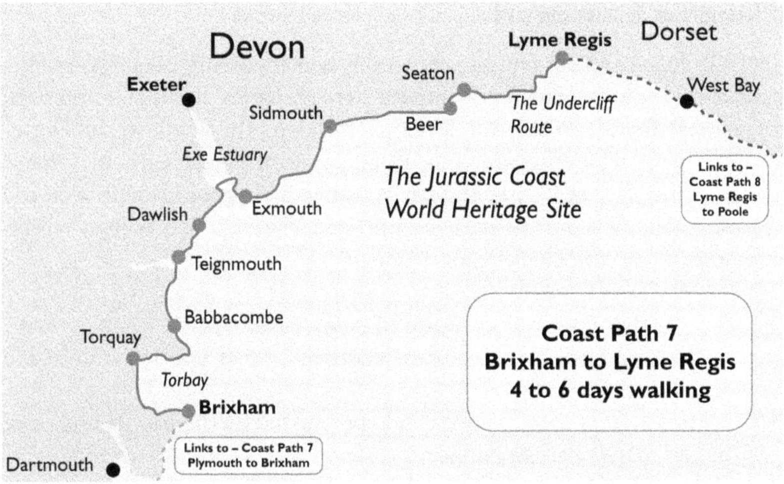

Figure 2.1 The Jurassic Coast

(*Source:* Encounter Walking Holidays Ltd, www.encounterwalkingholidays.com, with permission.)

Heritage site aims and objectives

The Jurassic Coast has been successful in achieving the coordinated and integrated management of an important heritage site for conservation, education and public enjoyment, as a result of the following strategies:

- An overall Coastal Management Plan and full institutional policy support.
- Involvement of over 100 stakeholder organisations in the Dorset Coastal Forum.
- Dedicated, full-time WHS staff, with a range of disciplines, coordinating activities supported by volunteers.
- Stimulation of local initiatives and the encouragement of innovation.
- Use of a strong market friendly brand and its application across a range of media.
- Community initiatives including 'gateway' towns and villages.
- Use of voluntary codes of practice to manage recreational activities.
- Establishment of Dorset Maritime Archaeological Record and the Marine Biodiversity Database.
- Founding a WHC Trust as a charity and independent body to raise funds and grow a Friends organisation.

Planning framework

In 1995 Dorset Coast Forum was established involving over 100 public, private and voluntary sector organisations to prepare a sustainable approach to management, land use and development along the whole of the Dorset coast. This is widely regarded as an effective strategic body. In 1999 it produced the Dorset Coast Strategy, an integrated approach with a clearly stated vision, agreed principles and priorities. The Coastal Forums' vision is that, by 2050, the Dorset coast will be :

- A coast that is at least as beautiful as it is now.
- A thriving and diverse coastal economy based upon a sustainable development approach.
- A coast that is used, enjoyed and appreciated by the people of Dorset and an active visiting public.
- A world-leading area in coastal management.

The Dorset coastal planning framework is mirrored by a similar approach for East Devon.

On a more local level, there are seven local authorities with responsibility for the coast in the WHS-designated area and three different shoreline management plans. Many local initiatives are now taking place, including:

- Developing a marine industry based at Portland including the 2012 Olympic Sailing Academy.
- The first integrated archaeology database in the UK.
- Bans on pair trawling within 12 miles of coast.
- Beach water quality improvement.
- Christchurch Harbour Group for zoning water recreational uses.
- A renewable energy scheme.
- Voluntary codes for fossil collecting and recreational activities.
- Completion of the South West Coast Path for walkers.
- A data collation working group.
- Eight voluntary bathing-only zones.
- The production of 18 Topic Papers to inform strategic planning, including Recreation and Tourism.

Organization structure

The WHS is managed by a full-time Executive working to the agreed management plan. This is a key document and concentrates on four areas, adopting UNESCO guidelines: conservation, education, science and access. The Executive consists of six full-time staff: Team Leader, Earth Sciences Advisor and Administrator, Education Officer, Visitor Management Officer and Projects Officer. This team is guided by a Steering Committee with an independent Chair and five specialist topic groups: Science and Conservation Advisory Group, Education Working Group, Gateway Towns Group, Tourism Working Group and Museums Working Group.

There is a separate World Heritage Coast Trust which is a fundraising body that promotes understanding and awareness of the conservation of the WHS. Core funding is provided by Dorset and Devon County Councils.

Tourism

Tourism is the area's predominant industry employing 38,000 people (65% work in the coastal zone) and generating over £830 million. The coastal zone attracts 16.5 million visitors each year, 1 million of which are overseas tourists. Of the total visitors, 3.5 million are overnight tourists and 13 million are day visitors.

The chains of ten coastal interpretive centres and 17 local tourist information centres (TICs) are vital to visitor management. Local libraries and public offices are also used as Tourist Information Points. Tourists are encouraged with the stimulation of a range of high-quality publications and a 'new generation' of quality postcards, along with a widespread use of the Jurassic Coast brand and imagery.

The main natural attractions are the beaches and coastal landscapes. Lyme Regis is famous for its connections with Jane Austen, Mary Anning and its Cobb. Chesil Beach attracts visitors for sea angling. The Isle of Purbeck, Portland Bill and Hengistbury Head are main attractors. An estimated 1 million people a year walk the South West Coast Path between Poole and Lyme Regis. There are several private estates with nationally recognised coastal features that have developed a tourism business, notably the Weld Estate at Lulworth and the Encombe Estate at Kimmeridge. There are nine major visitor attractions in the coastal zone including the Abbotsbury Swannery (106,000 visitors), Swanage Steam Railway (140,000 visitors) and the Christchurch Priory (160,000 visitors), as well as numerous smaller attractions.

The Charmouth Heritage Coast Centre

One of the chain of interpretive centres along the Jurassic Coast and established in 1985, The Charmouth Heritage Coast Centre is widely regarded as a success story in encouraging the safe and sustainable collecting of Jurassic fossils on local beaches. The Centre provides an interpretive exhibition (i.e. interactive computers, video microscopes, etc) and information on fossils, fossil hunting and the coastline in general (also with the use of a theatre showing informative films). Various interpretive walks, tours and events are run throughout the year, focusing on four main themes: marine activities, fossil hunting walks, fossil Road Shows, and sea dragons and volcanoes.

Public events are integrated with a programme of schools and field study products, charged on a commercial basis. The full-time warden is supported by two summer placement wardens and a volunteer group. The Centre is run by a charitable trust, The Fine Education Foundation. Income is derived from a range of sources: entry to the film theatre, walks, tours and events (for example fossil weekends for up to 40 people), sales of publications, educational activities, research, Friends (donations) and rent from the fossil shop (on the ground floor).

The Friends organisation is regarded as the lifeblood of the Centre. Most volunteers are drawn from within the local community and are involved in all aspects of the Centre's operation. The annual Friends donation is £5 per annum, and there are two categories of Friends: (a) Associate Member (passive); or (b) Working Member. The average annual visitation is 65,000–70,000 including an estimated 5,000 school children as part of formal educational groups.

Sustainable tourism

The Jurassic Coast offers a wide range of local tourism services and products while piloting the Green Tourism accreditation in the WHS. Specific initiatives designed to promote good practice and to achieve sustainable outcomes at key locations include:

- Poole Harbour Aquatic Management and Zoning Plan
- Christchurch Harbour Management Plan
- Durleston Marine Project
- The Limpet Protection Zone
- Purbeck Marine Wildlife Reserve.

There are also a wide range of voluntary conservation areas and codes of conduct for different recreational activities. The work of the sea fisheries and conservation teams integrates closely with the tourism effort, including a number of food initiatives:

- The Lyme Bay Reefs Project, run by the Devon Wildlife Trust
- The Weymouth and Portland Fish Festival
- The Seafood from the Dorset Coast Project
- The Fish and Chip Wrapper Project (designed to increase awareness of conservation using wrappers as interpretive media).

Although 80% of visitors travel by car, Coastlinx has proved to be a successful sustainable transport initiative, receiving several sustainable tourism awards. This is a public bus service linking all key villages and attractions along the WHS coast and integrating with other transport systems (both mainline rail services and local transport providers such as the Seaton Tramway). Operated jointly by Dorset and Devon County Councils, it offers a jump-on–jump-off service all year round. An unlimited travel ticket is available per day at £5 per adult and £12.50 for a family of four. There are specially converted double-deckers with upper floor viewing windows. The service is promoted widely with a high-quality website, and leaflets and timetables available throughout the WHS in TICS, local shops and libraries.

In addition to Coastlinx, other sustainable transport initiatives include: The Purbeck Cycle Way, Swanage Railway Park and Ride, and the Weymouth Hopper Bus.

Heritage site challenges

Some key issues associated with tourism and maintaining the local environment include:

- Inappropriate development pressures and land use on accessible, attractive, coastline. These are particularly fierce in the east of the designated area.
- Managing water quality with the increasing population, both permanent and temporary (tourists).
- Increasing range and variety of water-based recreational activities, notably sub-aqua, jet-skis and power-boating.
- Stimulating local product innovation and community buy-in.
- Beach cleanliness and tourist litter.
- Meeting increasingly discerning visitor expectations, especially in terms of interpretation (post *Coast* series on BBC TV), accommodation, and local food and drink.
- Visitor management, including car parking and traffic congestion, especially in hotspots such as Lulworth Cove, Shell Bay, Charmouth and Lyme Regis.
- Maintaining the quality of all tourism products.
- Generating revenues to keep heritage projects going.

Conclusion

The Jurassic Coast WHS initiative is an excellent example of how different authorities, stakeholders, academics and local communities can collaborate to develop an integrated coastal strategy that will deliver economic benefits while at the same time protecting miles of Jurassic coast incorporating England's only natural WHS. Regardless of whether communities work via a cluster or forum, the procedures, principles and practices of collaborative economic development remain the same. Instead of a fragmented approach, the widespread use of the Jurassic Coast brand and imagery has achieved a triple bottom line benefit for all concerned. The author argues that this successful collaborative and consultative initiative would probably not have been achieved via a single institutional structure.

References and further reading

Chan Kim, W. and Mauborgne, R. (2015). '8 key points of blue ocean strategy'. www.blueoceanstrategy.com/sail-toward-your-own-blue-ocean/8-key-points-of-bos-ebook.pdf (accessed: April 2017).

Chan Kim, W. and Mauborgne, R. (2015). 'Blue ocean strategy'. INSEAD Blue Ocean Strategy Institute. https://centres.insead.edu/blue-ocean-strategy/documents/e-ibosi2015.pdf (accessed: April 2017).

Dalhousie University (2010). 'Marine biodiversity strongly linked to ocean temperature'. www.sciencedaily.com/releases/2010/07/100728131707.htm (accessed: May 2017).

European Commission (2010). 'Towards an integrated maritime policy for better governance in the Mediterranean'. http://eur-lex.europa.eu/legal-content/EN/TXT/?uri=CELEX%3A52009DC0466 (accessed: April 2017).

European Commission (2012). 'Blue growth: Opportunities for marine and maritime sustainable growth'. https://ec.europa.eu/maritimeaffairs/content/blue-growth-opportunities-marine-and-maritime-sustainable-growth_en (accessed: April 2017).

European Commission (2012). 'Integrated maritime policy'. https://ec.europa.eu/maritimeaffairs/policy_en (accessed: April 2017).

OECD (Organisation for Economic Co-operation and Development) (2016). *The Ocean Economy in 2030*. Paris: OECD Publishing.

Stevens & Associates (2006). 'A review of relevant experience and best practice in sustainable tourism in the coastal and marine environment'. Scottish Natural Heritage Commissioned Report. www.snh.org.uk/pdfs/strategy/natparks/CMNP/advice/Final%20report%20-%20CMNP%20sustainable%20tourism%20-%20February%202006.pdf (accessed: April 2017).

Tittensor, D.P., Mora, C., Jetz, W., Lotze, H.K., et al. (2010). 'Global patterns and predictors of marine biodiversity across taxa'. *Nature*, 466: 1098–1105.

United Nations Educational, Scientific and Cultural Organization and Dorset and East Devon Coast (2014). 'Dorset and East Devon Coast World Heritage Site: Management plan, 2014–2019'. http://jurassiccoast.org/wp-content/uploads/2015/10/Jurassic-Coast-World-Heritage-Site-Management-Plan-2014-%E2%80%93-2019-Approved.pdf (accessed: April 2017).

United Nations Educational, Scientific and Cultural Organization and World Heritage Convention (2009). 'An economic, social and cultural impact study of the Jurassic

Coast: A summary of findings'. http://jurassiccoast.org/wp-content/uploads/2015/10/Economic-Impact-Study-2009.pdf (accessed: April 2017).

United Nations Environment Programme (2012). 'Green economy in a blue world synthesis report'. https://sustainabledevelopment.un.org/index.php?page=view&type=400&nr=558&menu=1515 (accessed: April 2017).

World Wildlife Fund (2015). 'Reviving the ocean economy: The case for action'. Washington, DC: WWF. www.worldwildlife.org/publications/reviving-the-oceans-economy-the-case-for-action-2015 (accessed: April 2017).

Chapter 3

Management strategies within a marine spatial planning framework

With each activity competing for the same space, a typical conflict of interest exists among different economic user groups. The range of international agreements dealing with different aspects in the ocean realm has contributed to international norm-setting for ocean governance, but implementation at the national and regional levels continues to be a challenge. Many of the marine scientific research and maritime safety and security issues are beyond the capacity of individual states to tackle alone. Growing dependency on ocean use for food security has the potential for conflict among and within nation-states. In addition, growth in the intensity of ocean use is impinging on the carrying capacity of the marine environment.

Marine protected areas (MPAs)

Our seas are vital for our existence. They support 80% of the world's biodiversity, supply more than 50% of the oxygen we breathe, and strongly influence the world's climate. They also provide us with more than US$21 trillion worth of ecosystem goods and services, such as nutrient cycling, raw materials, food and tourism. Despite this value, they face a vitriolic combination of threats, including pollution, habitat loss, climate change and overfishing. These anthropogenic pressures have depleted populations of culturally and economically important fish stocks and reduced the structural complexity of various communities across a rich range of habitats, species and trophic levels. Due to the broad scope and multifaceted nature of the stressors, there has been growing recognition of the importance of an integrated ecosystem approach to the management of the ocean. At the centre of such an approach is the use of marine protected areas (MPAs): parts of the ocean that are protected from at least some extractive or damaging activities (Worm et al. 2006).

Numbers of MPAs are expanding rapidly (Gray and Campbell 2008). Today there are more than 5,000 worldwide, but only about 0.78% of the world's oceans have MPA designation, while just 0.02% are afforded full protection from all damaging activities (WCPA/IUCN 2007). For better or worse, the conservation community and resource management authorities frequently measure their progress by metrics of the number and geographic coverage of MPAs. Indeed, a number of international agreements, such as the Convention on Biological Diversity's 2011 Aichi Biodiversity Targets, set a bar for critical conservation at 10% coverage by MPAs of the ocean within national jurisdictions and within bio-geographic regions. Several scientific analyses, however, have indicated that having 20–30% of marine areas in reserves is the minimum necessary to safeguard marine life. Despite such warnings, we are now watching a very slow race to achieve 10% of marine areas in some form of protected zone (at the current rate, this might take until 2100).

At the 3rd International Marine Protected Areas Congress (IMPAC3) meeting in France in October 2013, the IUCN World Commission on Protected Areas (WCPA) and UNEP World Conservation Monitoring Centre (WCMC) announced 'The Official MPA Map' at www.ProtectPlanetOcean.org based on the latest World Database of Protected Areas (WDPA). This analysis suggests that 2.8% of the total ocean is protected, with about 10% of territorial seas (0–12 nautical miles), 5% of exclusive economic zones (EEZ) (12–200 nautical miles), and 0.14% of the high seas (beyond 200 nautical miles) within MPAs. This serves as a good upper-limit for the total amount of global MPA coverage. However, the actual amount of coverage by parks that have permanent protection is a little less, and the coverage of fully no-take areas and other strongly protected and well-enforced areas is even less still.

MPAs are fully protected from all damaging impacts and managed by governments, whereas marine reserves are partially protected and may be managed privately or by provincial or municipal authorities. These especially can have a number of benefits. The removal of fishing pressure has been shown to increase average size, diversity, abundance and biomass of invertebrates and fish within protected areas.

There is also evidence that these benefits may have positive consequences for fisheries outside the MPA. This can occur through two mechanisms: net export of pelagic eggs and larvae (recruitment effect) and net migration of juveniles and adults across protected area boundaries (spillover effect). Overall, it is clear that there is solid evidence to suggest that MPAs and marine reserves can be valuable management tools for the restoration and maintenance of marine ecosystems and thereby more appealing to nature-based tourists seeking a pristine tourism experience.

However, in many cases MPAs fail to actually protect. Inadequate long-term funding and widespread management failure have resulted in unenforceable and ineffectual 'paper parks'. As an example, too many authorities allow fishing rights in parks in order to win political support. In a worldwide study of MPAs, less than 16% of protected area managers felt they had adequate funding for effective conservation. Most MPAs are therefore failing and, as such, there has been increasing interest in alternative mechanisms for financing and management.

South Africa's MPAs: Lack of investment

South Africa's 24 MPAs are critical for the health of future fish stock, but their significance is not fully appreciated. From coastal residents and fishers through to government prosecutors and judges, a lack of understanding of the importance of these areas continues to hamper efforts to manage MPAs effectively. These are among the findings detailed in the World Wide Fund (WWF) South Africa's latest marine assessment, titled 'The State of Management of South Africa's Marine Protected Areas 2014'. The report follows a commitment by the government to develop an 'oceans economy' as part of the Operation Phakisa initiative.

Operation Phakisa is a South African government initiative designed to fast track the implementation of solutions on critical development issues. This is a unique initiative to address issues highlighted in the National Development Plan (NDP) 2030, such as poverty, unemployment and inequality. Operation Phakisa is an innovative and pioneering approach to translate detailed plans into concrete results through dedicated delivery and collaboration. *Phakisa* means 'hurry up' in Sesotho and the application of this methodology highlights the government's urgency to deliver. It plays a crucial role in accelerating the delivery of some of the development priorities. Through Operation Phakisa, the government aims to implement priority programs better, faster and more effectively.

Similar to land-based parks and reserves, MPAs are created to protect important biodiversity assets from certain activities such as fishing, mining and other developments. The WWF-SA report highlights the key challenges and practical solutions which need to be addressed in order to advance the country's existing network of 24 MPAs, among them Table Mountain National Park MPA in Cape Town and the Bird Island MPA near Port Elizabeth.

With many of the local marine resources overexploited, and alarmingly 47% of marine and coastal habitat types considered threatened, the need for effective protection has never been clearer. Recognised as a key tool in fisheries management and biodiversity conservation, MPAs form the foundation of marine conservation and also meet a wide range of human needs. Fisheries in the vicinity of MPAs can recover rapidly, possibly without negative consequences for fishers due to increased travel distances and reduction of catch rates. This recovery is best explained by a succession of spillover of adult fish from the MPA followed by a recruitment boost caused by increased larval dispersal. For example, a study conducted in the Tsitsikamma MPA in the southern Cape revealed that the density of the commercially important red roman (*Chrysoblephus laticeps*) was approximately 42 times higher in the MPA than in nearby fishing grounds.

Following similar assessments of the management effectiveness of South Africa's MPAs in 2003 and 2009, a repeat assessment was conducted in 2013 to evaluate the progress made through the actions of national and provincial agencies involved in MPA management. Despite some important areas of improvement, overall, the report shows that many challenges identified in 2009 have persisted in 2013.

Marine conservation agreements (MCAs)

Terrestrial private protected areas safeguard millions of hectares of biologically significant habitat worldwide, but uptake on submerged lands has been limited, due primarily to the erroneous assumption that the oceans are part of the commons and cannot be owned or leased. In fact, commercial enterprises have been acquiring rights to marine and coastal areas for centuries: every year they spend billions of dollars acquiring rights for fisheries, aggregates, energy, marinas and other uses. Increasingly, by using marine conservation agreements (MCAs), organizations like The Nature Conservancy (TNC) have begun to acquire similar rights for purposes that safeguard marine resources and benefit local communities.

TNC and Conservation International define MCAs as any formal or informal understanding between two or more parties in which the parties obligate themselves, for an exchange of benefits, to take certain actions, refrain from certain actions, or transfer certain rights and responsibilities to achieve agreed-upon ocean or coastal conservation goals (The Nature Conservancy and Conservation International 2012).

Although MCAs are a promising tool in international marine conservation, their use to date has been limited; and the reasons for this are twofold. First, the role of MCAs in conjunction with other marine conservation strategies is not well defined or documented, leading to a lack of awareness, understanding and common base of practice within the marine conservation community. Second, different nations have disparate legal and policy frameworks, ownership practices and management regimes, and it is necessary to understand these to determine the feasibility of MCAs in a specific location. However, this limitation is being addressed by TNC, who are developing a wide-ranging analysis of the role and feasibility of MCAs in key regions worldwide (Hale 2009).

This analysis examines one such region – the western Indian Ocean – and documents findings from an analysis of the ocean and coastal legal frameworks in Kenya, Tanzania, Mozambique, Madagascar, Seychelles, Comores, Mauritius, and South Africa. An exhaustive synthesis of the laws, policies and practices relating to use and management of coastal resources in the region reveals that MCAs are a feasible strategy in four of the five countries under consideration, and identifies suitable conservation partners. Through an additional analysis of the responses of key opinion leaders in the region, ten MCA projects are identified and summarized and one detailed case study developed. Finally, the report identifies two limitations that may caveat interpretation of findings and suggests additional focused assessments to satisfy information needs.

MCA or MPA?

Given the broad nature of definitions, it can be arduous to separate MCAs from their MPA cousins. As one attendee at a December 2009 TNC workshop opined: when is something an MCA and when is it not? Does a specific public process and agreement to MPA zoning with the expectation that the public will receive ecological benefits in the future count as an MCA? If so, practically everything is an MCA. The International Union for Conservation of Nature (IUCN) defines MPAs as places in the marine environment with:

> clearly defined geographical space[s], recognized, dedicated and managed, through legal or other effective means, to achieve the long-term conservation of nature with associated ecosystem services and cultural values.
>
> (IUCN/WCPA 2008)

As such, an MPA is a designation applied to a place; and an MCA is an agreement to act or refrain from acting (The Nature Conservancy 2010). An MCA can help parties assent to the formation of an MPA (The Nature Conservancy 2010). It can compel parties to act in certain ways in certain places, causing areas to operate like MPAs.

Marine spatial plans (MSPs)

Demand for outputs (goods and services such as food and energy) usually exceeds the capacity of marine areas to meet all of the demands simultaneously. Marine resources are 'common property resources' with open or free access to users. Free access often, if not always, leads to excessive use of marine resources, for example overfishing, and eventual exhaustion of the resources. The fact that not all of the outputs from marine areas, especially natural services such as wildlife habitat and nutrient cycling, can be expressed in monetary terms, means that markets cannot perform the allocation tasks. Some public process must be used to decide what mix of outputs from the marine area will be produced over time and space. That process is marine spatial planning.

The advantages of marine spatial plans (MSPs) include:

- Allow a more strategic, pro-active approach to planning.
- Promote rational use of marine space and resources and sustainable development of maritime regions.
- Involve all stakeholders and map their interests and uses.
- Coordinate sectors/users to achieve agreed-upon goals and objectives.
- Seek to balance economic use and conservation.
- Enable early identification of potential conflicts before considerable investment or damage has occurred.
- Provide greater certainty in acceptable locations for different types of use.
- Improve understanding and consideration of the cumulative effects of different activities.

Defining marine spatial planning

Marine spatial planning is a public process of analyzing and allocating the spatial and temporal distribution of human activities in marine areas to achieve ecological, economic, and social objectives that usually have been specified through a political process. Characteristics of marine spatial planning include ecosystem-based, area-based, integrated, adaptive, strategic and participatory. Marine spatial planning is not an end in itself, but a practical way to create and establish a more rational use of marine space and the interactions between its uses. It balances demands for development with the need to protect the environment, and it delivers social and economic outcomes in an open and planned way. In order to clarify the definition of MPAs in contrast to other marine-managed areas, the IUCN uses the following:

> A clearly defined geographical space, recognized, dedicated, and managed through legal or other effective means, to achieve the long-term conservation of nature with associated ecosystem services and cultural values.
>
> (IUCN/WCPA 2008)

In order to better define the term 'marine protected area' for the purpose of building and implementing a national MPA system, the MPA Centre uses criteria based on the official definition of a marine protected area in the MPA Executive Order 13158:

> [. . .] any area of the marine environment that has been reserved by federal, state, territorial, tribal, or local laws or regulations to provide lasting protection for part or all of the natural and cultural resources therein.

Both definitions clearly indicate the importance of permanence and place-based management. For this reason, the MPAtlas analysis pulled out a number of fisheries management zones that are temporary closures (i.e. the promise of permanence is not met) and broad regional management zones without clear place-based affinity and management (i.e. a single gear closure along the entire US East Coast does not equal a marine protected area). With a pass through the database, the MPAtlas removed about 2% of the global marine area covered by these types of non-marine-protected-area, and fisheries management areas that are not also covered by other protected areas.

What marine spatial planning is NOT

Marine spatial planning is not a substitute for single-sector planning and management. Strategic and operational plans for fisheries, transportation, energy, recreation, and conservation, for example, will continue even when integrated MSP is put into practice. Integrated MSP can provide a guide to single-sector management that should increase compatibilities and reduce conflicts across sectors, balance development and conservation interests, increase management effectiveness and efficiency, and address the cumulative effects of multiple human uses of the same marine space.

Marine spatial planning is not a one-time plan. The context for planning is constantly changing. Science contributes new knowledge. Monitoring and evaluation adds new information about the effectiveness, efficiency and equity of alternative management actions. Technology improves social, economic and political conditions that all change over time. Plans should be updated periodically to reflect these changing conditions. Marine spatial planning is not only conservation planning. While a network of MPAs might be one outcome of MSP, it seeks to balance economic development and environmental conservation, and not focus on only the goals of conservation or protection.

Marine spatial planning is not ocean zoning. Marine space has been zoned for individual human uses for decades, if not longer. Fisheries have been opened or closed in particular areas or zones. Coastal management plans have zoned areas suitable for coastal recreational activity. Marine transportation has been managed within designated lanes or zones especially in intensively used areas. Rights to explore or exploit energy or mineral resources have been leased on an area basis. MPAs have been designated in many places in the world. However, these zones and others have usually been planned on a single-sector basis without integrated planning.

What are the outputs of marine spatial planning?

The marine spatial planning process results in a comprehensive plan or vision for a marine area. Marine spatial planning is one element of ocean or sea use management; zoning plans and regulations are one of a set of management actions for implementing marine spatial planning. Zoning plans can then guide the granting or denial of individual permits for the use of marine space.

Why marine spatial 'management' and not only 'planning'?

Planning is only one element of the marine spatial management process. This process includes additional elements of implementation, enforcement, monitoring, evaluation, research, public participation and financing, all of which must be present to carry out effective management over time. Certainly conflicts of interest within a local maritime domain can be substantially reduced if all coastal and ocean stakeholders residing and working within that domain arrange a consultative workshop to map their domain according to appropriate development criteria onshore, near shore and offshore within a local marine spatial plan framework. Thereafter, coastal and ocean developments confined to coastal and ocean development zones should be evaluated after undergoing a thorough risk assessment and cost benefit analysis to ensure that the proposed development does not impact negatively on existing developments of much higher value.

Coastcare

In South Africa, The Department of Environmental Affairs and Tourism provides national leadership for promoting sustainable coastal development. This is primarily achieved through Coastcare, a partnership involving the private and public sectors. The policy guiding this joint initiative is presented in the White Paper for Sustainable Coastal Development in South Africa.

Coastcare aims for:

- Coastal economic development that makes the best use of available resources.
- Coastal development that promotes social equity through improved livelihoods of poor coastal communities.
- A healthy coastal environment for the benefit of current and future generations.

Coastcare provides financial and technical assistance for:

- Coastal development projects
- Institutional capacity building of coastal management organisations
- Legal development to support policy
- Awareness education and training initiatives
- Coastal resource planning
- Applied research
- Coastal information management projects.

Figure 3.1 illustrates a typical coastal protection zone managed by the Coastcare partnership program implemented by the Department of Environmental Affairs and Tourism. Coastal protection zones act as a buffer when terrestrial protected areas interface with MPAs.

Management strategies for marine spatial planning

Within a marine spatial planning framework, management strategies can be divided into four main categories: regulatory; physical; economic; and educational.

Regulatory management strategies

Rules and regulations are commonly utilized to restrict visitor actions, access, times and numbers, and are usually posted on signs, notices and written material. Enforcement is usually undertaken by police, park rangers or other agents of the management authority. A system of progressively harsher punishment is most common, for example, warning at the first offence, banning from the area at a second offence, and eventually fines and imprisonment in extreme cases of misconduct, such as harassment or killing of protected wildlife. Examples of regulations

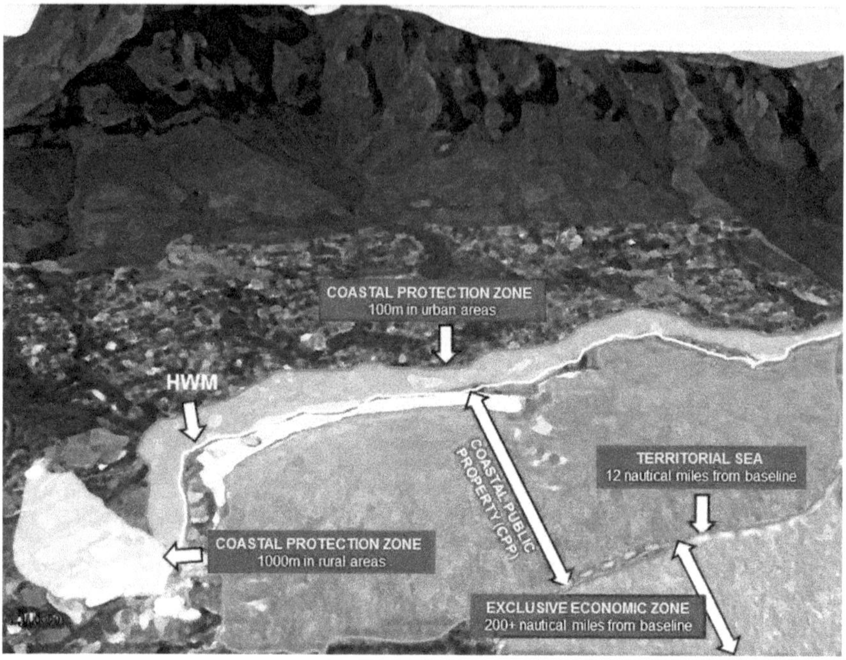

Figure 3.1 Coastal protection zone (HWM is high water mark)

(*Source:* Department of Environmental Affairs, South Africa, with permission.)

commonly used to manage tourists are visitor number restrictions, limits on times, locations and noise levels, specification of types of visitor activity permitted, spatial and temporal zoning, restrictions on the type of equipment permitted, speed limits, permits, lease and licence requirements, and codes of practice. The purpose of regulatory practices is threefold: first, to protect the safety of the tourist; second, to reduce conflicts between tourists; and third, to protect the marine environment from negative impacts due to inappropriate tourist behaviour.

Physical management strategies

Physical approaches to management are those human-made structures that control human activity by restricting the movement or type of activity which can be undertaken. A typical marine example is the construction of a boardwalk across a wetland. This physical structure directs and facilitates the movement of tourists and reduces the negative impacts that could be caused by their walking through sensitive areas. Due to the difficulties involved with erecting structures in the marine environment, physical controls are not a particularly common system in controlling marine tourism. In a number of situations, physical structures have been successfully used to control tourists in the sea, for example, the use of mooring buoys to reduce anchor damage on coral reefs. Additional examples include the use of glass-bottom boats, semi-submersibles, self-guided underwater trails and beach bicycle pathways.

Economic management strategies

Economic strategies use price as incentives or disincentives to modify people's behaviour. An example of this type of strategy is the use of higher entry fees to facilities during peak use times in an attempt to spread visiting. Imposing fines for littering, taking undersize fish or other inappropriate behaviour is another example of a regulation combined with an economic disincentive. Discounts on access fees to a marine park could be provided if groups undertake a clean-up project, or assist with research, during their visit.

Educational management strategies

The goals of education-based strategies are to reduce the incidence of inappropriate tourist behaviour by encouraging voluntary behaviour change, and to increase visitor enjoyment and understanding. Many researchers, therefore, view education as a potential 'win–win' situation for both the marine environment and the tourist. The planning and implementation of an effective educational program for tourists is more difficult than other strategies due to a number of factors, for example, the diverse characteristics of tourist groups such as different group size, age and educational attainment. Lack of knowledge regarding the marine environment and a lack of suitably trained and qualified personnel can also complicate efforts to implement an effective education program. These factors result in little use of education as a management strategy and the predominance of physical and regulatory approaches. However, a number of researchers argue that education is, or should be, a critical component of nature-based tourism experiences.

Responsible marine tourism

The overall goal of management strategies designed to control marine tourism is twofold: first, to protect the marine environment from detrimental impacts; and second, to provide for and to promote enjoyable tourist experiences. However, in order to accomplish this ambition, tourists must somehow be prompted to do more than simply 'have a good time'. Furthermore, and equally important, commercial operators must see their responsibilities as extending beyond making a profit. They must see their role as one of contributing to the health and viability of the resource upon which the livelihood of their business depends. If this can be achieved, the environment, upon which tourism is based, would actually benefit from tourism.

The fact is that many tourists are willing to give their time, money and labour in order to assist with nature conservation. The success of a management regime can, therefore, be measured in terms of its effectiveness in prompting the movement of the tourist experience towards these desired objectives. The first of these is changing tourists' behaviour and lifestyle so that their actions become more environmentally responsible, both during the tourism experience and, longer term, after the experience. The second is promoting tourist actions that directly contribute to the environment while they are visiting it; for example, through assisting with research projects, becoming involved with habitat restoration programs, removing litter, acting as volunteers for the management agency, or even helping to police the area so that other tourists' actions are not detrimental to the environment.

CASE STUDY: SOUTH AFRICA, PORT ELIZABETH AND ALGOA BAY

History

Port Elizabeth has many names: 'Algoa Bay', 'The Friendly City', 'The Windy City', and, officially, 'Nelson Mandela Bay'. It also has a long and colourful history, which includes the influences of many famous people. A warm and unassuming city, it boasts a rich heritage which has been formed by people from a multitude of different cultural groups. In 1576, Manuel de Mesquita Perestrelo named Port Elizabeth's bay 'Baia de Lagoa' because of the lagoon at the mouth of the Baakens River that flows through Port Elizabeth (See Figure 3.2). This has been adapted to the current name, 'Algoa Bay'. Algoa Bay was first settled by hunting and gathering people ancestral to the San at least 100,000 years ago.

A little over 2,000 years ago, agriculturalist populations ancestral to the Xhosa migrated into the region from the north, eventually displacing or assimilating the region's indigenes. For centuries, the area was simply marked on navigation charts as 'a landing place with fresh water'. The greater Nelson Mandela Bay metropolitan city incorporates the City of Port Elizabeth and the adjacent towns of Uitenhage and Despatch. Historically, Port Elizabeth is a maritime city. Bartholomeu Dias arrived in Port Elizabeth more than 500 years ago in September 1488, looking for a trade route to India and hoping to find a trace of Prester John, the benevolent ruler of a lost kingdom of great beauty and wealth.

Prester John, so the legend said, was a descendant of one of the three Magi and had founded a prosperous kingdom in an unknown location. Here, peace reigned and man lived in harmony with strange creatures alongside the fountain of youth. The legend of Prester John fired the popular imagination and spurred quests to faraway lands for several hundred years. By the time the legend started to fade, it had directly and indirectly encouraged Europe's explorers, missionaries, scholars and treasure hunters to set sail and discover the world beyond their own borders. Today, the only monument in the world commemorating this powerful influence can be found in Port Elizabeth. The memorial stands in the inner city, adjacent to the magnificent City Hall, and is dedicated to both the Portuguese

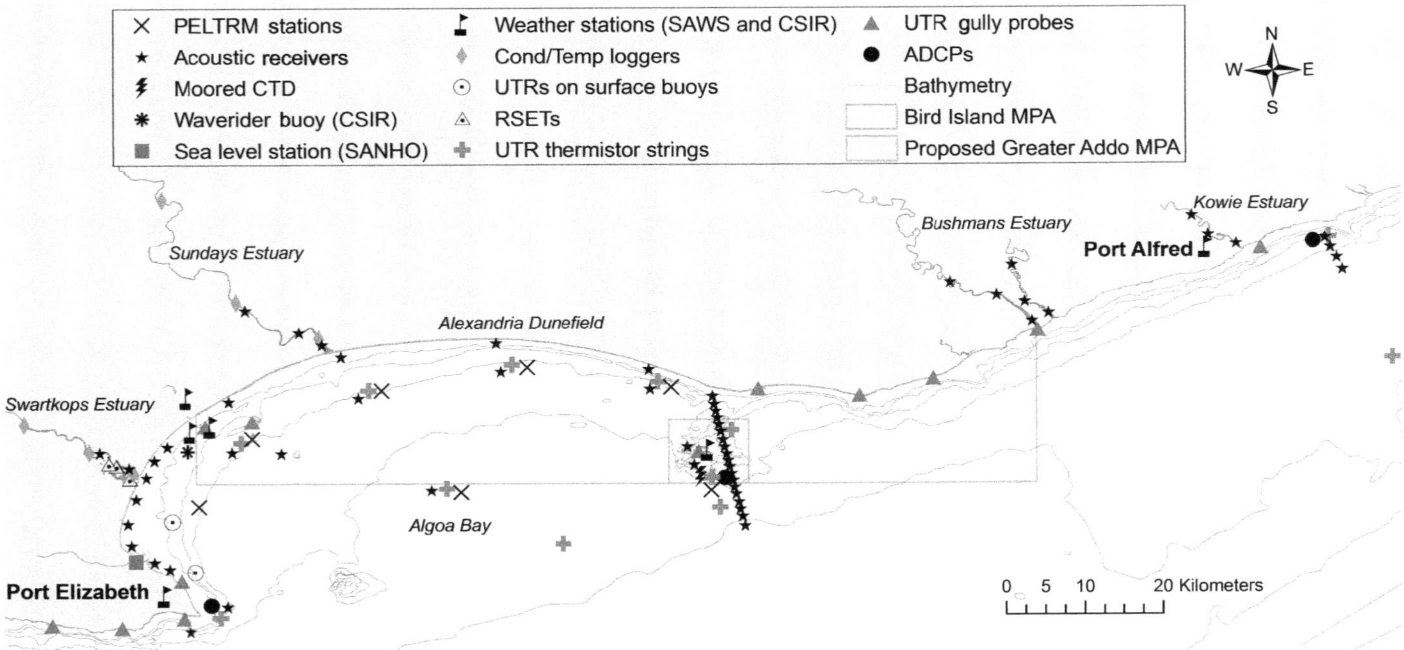

Figure 3.2 Algoa Bay

(*Source:* South African Environmental Observation Network (SAEON), with permission.)

explorers who came to the shores of Algoa Bay so many years ago and to Prester John. Perhaps this is the place that could demonstrate to the world that humankind, wildlife, marine life, commerce and industry, can all live in harmony in a unique coastal and marine tourism domain unlike any other in the world. It is in this area that the first meetings of the Khoisan, British, Dutch, German and Xhosa people took place.

Today, Port Elizabeth is still a melting pot for a myriad of races and nationalities – a true representative of the Rainbow Nation. Algoa Bay is considered by many to be a perfect bay for yachting, boat cruises, windsurfing and diving. The mix of colourful coral reefs and shipwrecks provide fascinating scuba-diving experiences. Hobie Beach was host to the world boardsailing championships on two occasions. Kings Beach is hailed as one of the country's safest beaches. Sardinia Bay is one of the most picturesque beaches in the country. Algoa Bay is a biodiversity hotspot in the marine environment. International divers frequently report observing more mammal and shark species per dive than in some of the world's best diving destinations. Port Elizabeth is surrounded by over a million hectares of diverse protected areas offering a choice of national parks, provincial game reserves, marine reserves, wilderness areas, private game reserves and hunting safaris. All the coastal assets of Nelson Mandela Bay add up to a potential world-class marine tourism destination.

Algoa Bay marine biodiversity hotspot and international 'Hope Spot'

World famous oceanographer, Dr Sylvia Earle, visited Port Elizabeth in April 2013 where she received an honorary doctorate from the Nelson Mandela Metropolitan University in recognition of her outstanding contribution to oceanic education and awareness. On 11 December 2014, Dr Sylvia Earle placed a plaque at the Bay World Marine Complex to declare Algoa Bay a global 'Hope Spot'.

Algoa Bay was identified as a superb Hope Spot because there are clear plans to promote education, science, tourism and a sensible development of conservation and sustainability. The Sustainable Seas Trust (SST) arranged the South African Maritime Safety Authority (SAMSA) SEA Pledge Tour of Hope which was launched by Dr Sylvia Earle in April 2013. This led to many recommendations and an enthusiasm around South Africa for Hope Spots. Furthermore, the seating of the South African International Maritime Institute (SAIMI) and Chair of Law of the Sea and Development in Africa at the Nelson Mandela Metropolitan University (NMMU) will ensure that the university is fully equipped and prepared to play a key role in marine scientific research, ocean governance and maritime sector development in Africa.

Algoa Bay underwater museum

Apart from its substantial natural assets, Algoa Bay is also custodian of important cultural assets such as over 300 shipwrecks that have been documented (Morris 2005). Shipwrecks are a fascinating and exciting consequence of man's desire to migrate and trade the waterways of the world.

For maritime archaeologists, a sunken vessel presents a time capsule that makes research and preservation of these marine disasters both rewarding and stimulating. South Africa has a rugged and dangerous coastline. Its position on the trade routes between Europe and the East has resulted in over 2,400 vessels sinking, grounding, being scuttled or abandoned in its waters. Wrecked ships that are over 60 years old are protected in South Africa by the National Heritage Resources Act (25 of 1999) and may not be disturbed in any way except in terms of a permit issued by the South African Heritage Resources Agency (SAHRA). Historical dive sites, however, may be visited and enjoyed provided no artifacts are damaged or taken. To date the heritage value of these shipwrecks has never been calculated.

Ports and waterfront development

A study, compiled for the Port Elizabeth Regional Chamber of Commerce (Olver 2008), revealed that 92,586 jobs could be created in Nelson Mandela Bay during the 16-year construction period of a waterfront in the port of Port Elizabeth and close to 120,000 when the development was fully operational in 2030. Furthermore, the study showed that during the construction period a total of R12.7 billion would be spent directly in Port Elizabeth, resulting in a total contribution to gross geographic product (GGP) of R18.3 billion. The report stated that the scale of the project is 'significant' adding that if Port Elizabeth's gross domestic product (GDP) grows at six per cent a year then in 2030 'the waterfront and related developments will make up 32% of the GDP of Port Elizabeth'.

The development is also projected to attract an additional 46,000 foreign and 220,000 domestic tourists each year. The report concluded that the impact figures 'confirm the strategic importance of the waterfront development and underlines the critical importance of unlocking the land to the southern side of the port'.

The proposed waterfront development in the port of Port Elizabeth will include an attractive marina and other leisure tourism facilities which will greatly enhance Port Elizabeth as a cruise-line destination. City/port interfaces are complex territories where the inherent competition and complementarity between city and port play out in the face of limited spatial resources. The search for the right balance calls for solutions that guarantee a good spatial and functional mix capable of transforming and rejuvenating not only the city/port interface but also the entire territory of the port city.

Algoa Bay: A unique habitat

Algoa Bay, situated on the East Cape coast of South Africa, is known for its natural beauty and biodiversity. The bay incorporates two island groups and is home to the largest population of African penguins. It is a protected refuge to calving whales, large pods of bottlenose dolphin and humpback dolphins, and visiting orcas are not an uncommon sight. The annual migration of millions of sardine is announced and celebrated by blue-eyed Cape Gannets, game fish, dolphins and sharks alike.

The abundance is overwhelming and rare: hence the area is recognized as one of the world's temperate biodiversity hotspots, and declared a Mission Blue Hope Spot. Algoa Bay is also home to a wide variety of sharks, some residential such as ragged tooth, pyjama and leopard shark; others are frequent and regular visitors. The area is a nursing ground to both ragged tooth, great white and smooth hammerhead sharks. Sharks are both prey and predator and they fulfil an essential and meaningful role in the functioning of oceans as a whole: they are indicative of ocean life.

Algoa Bay is unique in the following ways:

- Algoa Bay is a declared marine biodiversity hotspot. Located at the conjuncture of two major oceanic systems, the Agulhas and the Benguela currents, the bay has a climate and oceanic conditions favourable for a large number of marine species.
- The islands in the bay host half of the global population of the iconic but endangered African penguins, and 70% of the vulnerable Cape gannets. These seabirds are decreasing dramatically in the rest of South Africa but are stable in Algoa Bay. Many other endemic seabird species breed in the bay, which led Bird Life International to declare the bay a marine important Bird Area.
- Algoa Bay is the only bay in South Africa containing two ports, i.e. a deepwater industrial port and a commercial port destined for a waterfront development and marina.
- Algoa Bay is one of the few bays in the world surrounded by over one million hectares of diverse land-based protected areas, i.e. national parks, provincial nature reserves, private game reserves and wilderness areas.

- The Greater Addo Elephant National Park is the only national park in the world where the 'Big Seven' land-based and marine-based species can be observed in one park, i.e. lions, leopards, rhino, buffalo, elephant, great white sharks and cetaceans (whales and dolphins).
- The bay hosts charismatic species of cetaceans, including killer whales, southern right and humpback whales, and four species of dolphins. The bay also contains large numbers of endemic invertebrates, seaweeds and fish (such as santer or red roman), and is a rare nursery ground for great white sharks, and southern right whales. The world famous sardine run also passes through the bay, feeding a whole suite of apex predators, and attracting international TV crews every year.
- Algoa Bay includes four estuaries that feed the productivity of the bay with nutrient input, while providing nursery areas for commercially important marine fish and invertebrates. The estuaries need to be protected from discharge of untreated waters that bring heavy metal and plastic pollution in the bay.
- The Alexandria Coastal Dune field is the largest coastal dune field in the southern hemisphere and a proposal has been submitted to UNESCO to declare the dune fields a mixed usage World Heritage Site.
- The bay is an underwater heritage museum containing over 300 historical shipwrecks. Remains of San settlements can be found and also evidence of Vasco da Gama's first trip around the Cape in 1488.
- Port Elizabeth is where the 1820 British settlers landed, bringing industrial skills and agricultural knowledge to the country. The British settlers introduced sport such as rugby, cricket, hockey and soccer to the country. Port Elizabeth has a legacy of firsts and is the place where many of the first international sporting events were held.
- The Port Elizabeth beachfront is acknowledged as one of the most beautiful and safest beachfronts in the country, especially for families, with Blue Flag beaches ideal for hosting many international events, such as the Iron Man, which generate large revenues for the city.
- Algoa Bay is considered to be a perfect bay for all sailing and water sport events, having hosted a number of world championship events. Scuba diving in the bay can be compared with some of the best dive sites in the world based on the diversity of marine species and spectacular colour of the hard and soft coral reefs.
- Algoa Bay is undoubtedly the most studied bay in Africa, contributing valuable marine scientific research data to the rest of the world since the relocation of the South African Environmental Observation Network (SAEON) to NMMU. The university is rapidly growing into a nerve centre for maritime research and innovation, policy development and

capacity building in Africa. Seated at NMMU are: the Coastal and Marine Research Centre, the South African International Maritime Institute, and the Chair of Law of the Sea and Development in Africa.

The Wildlife and Environmental Society of South Africa (WESSA) supported the initiative to launch 'Algoa Bay as the Bottlenose Dolphin Capital of the World'. This event took place on 18 April 2016 in the Algoa Bay Yacht Club attended by over 200 local residents and stakeholders. This proclamation is supported by marine scientific evidence based on records extending over a 20-year period of marine research studies, i.e. there are larger pods of bottlenose dolphin sighted more frequently in Algoa Bay than anywhere else in the world. WESSA also organizes coastal and marine education programs at schools and coastal clean-up events using volunteers from the local community.

Call for a sea ethic

An emerging literature on a 'sea ethic', 'ocean ethic' or 'new myth for the ocean' exists: examples include the Pew Oceans Commission, the Scientific Consensus Statement on marine ecosystem-based management, the work of Berkes (2012), and other books by environmental science writers. Almost all such calls refer to Aldo Leopold's (1949) 'land ethic' and/or the work of Rachel Carson (1951). These are the prophets of ecology, who dared to prescribe a right and wrong way to deal with the natural world. This appeal to the prophetic tradition attests to the discomfort of scientists who have left the safe shores of 'is' and ventured into the deep water of what 'should' be. A sea ethic can be summarized as working towards the flourishing of the entire biotic community, people, plants, animals, air, lands, waters and the life-sustaining flows between them.

To protect the considerable ocean resources and rich marine life in the bay, the Algoa Bay Hope Spot Steering Committee inspired by the SST South African initiative, and the Mission Blue global vision, prepared the following declaration as a contribution towards the development of a global sea ethic.

ALGOA BAY HOPE SPOT DECLARATION

We, the people of Algoa Bay, having witnessed the proclamation of Algoa Bay as a global Hope Spot, declare on this day that we will strive to sustain the marine biodiversity integrity of our maritime space by ensuring that it is protected and sustained in a healthy condition for all people and marine creatures to enjoy for future posterity;

Acknowledging that oceans are regarded as the 'last frontier' on earth, and that oceans decisively influence Earth's ecological balance, as well as world geopolitics, globalization, socio-economic balance and life on earth,

Accepting that oceans cover two thirds of the world's surface and as such the world's population, as well as all living things, are dependent on the oceans,

Recognizing that a failure to sustain the oceans in a healthy condition will lead to dire consequences for all humanity,

Conscious that the ocean space represents a unique socio-cultural entity that has offered sustenance and a distinct culture and way of life for centuries,

Convinced that the world oceans, seas and coastal areas are in a state of crisis, facing a greater array of problems and dangers than ever before in history such as toxic pollution from cities and fields, anthropogenic waste disposal, excessive nutrients and oil spills resulting in an increasing threat to living and non-living resources in the coastal and ocean environments thereby adversely impacting and fundamentally changing natural ecosystems, and even threatening human health,

Further convinced that marine life and vital coastal habitats are straining under the increasing pressure of deteriorating sea water quality and the cumulative effects of excessive human use and that the ability of marine ecosystems to produce the economic and ecological goods and services that are desired and needed, have been substantially reduced,

Aware that in some instances there has been a significant decline of ocean wildlife and even collapses of ocean ecosystems and with over half of the world's population living along coastal areas, there has been unprecedented commercial and residential overdevelopment,

Mindful that the condition of the oceans and the ecological state of the marine environment will ultimately determine the fate of planet earth, and that the four vital ecosystems for humankind and for all marine life forms, i.e. salt-marshes, mangroves, estuaries and coral reefs require the highest level of protection,

Accepting responsibility to work with boat-based tour operators to ensure that water-based recreational activities such as whale and dolphin watching are used as opportunities for marine education and awareness and in this regard to assist with sourcing funding to support these important coastal activities,

Committed to ensuring that onshore, near-shore and offshore recreational activities are appropriate for the marine environment and managed in accordance with integrated coastal zone management plans,

Endorsing that responsible ocean governance means integrated management combining the land and adjoining water in a single-unified framework for sustainable economic development to improve the quality of life of people; environmentally appropriate development; and equitable development in terms of inter-societal, inter-generational and international equity,

Convinced that the time has come to rethink how to manage oceans and recognizing that they are a key pillar for many countries contributing to their economic and social development which are vital in the fight against poverty, and that they can become more profitable, more sustainable, and meet the needs of a growing population without sacrificing the health of fragile ocean ecosystems,

Concerned that too many of these essential natural resources are being degraded by unsustainable use, putting the ecosystems services they provide, such as food security and climate regulations for instance, at risk,

Endorsing the concept of a 'Green Economy in a Blue World' which takes into account six different economic areas in the marine sector and provides recommendations on how to boost their potential by implementing green measures, the six economic areas being fisheries and aquaculture, marine transport, ocean nutrient pollution, marine-based renewable energy, coastal tourism, and deep-sea minerals,

Reasserting the overarching vision of the 2050 AIM Strategy to foster increased wealth creation from Africa's oceans and seas by developing a sustainable thriving blue economy in a secure and environmentally sustainable manner,

Agreeing that any development in the local maritime space should undergo a thorough risk assessment and cost benefit analysis,

Convinced that the maritime sector as a whole has much to gain by operating in a marine environment that is in a healthy condition; and which favours the market economy, private enterprise and free trade; and that serves to optimize its beneficial effects in a green blue economy that contributes significantly to the creation of wealth and employment,

Committed to supporting marine scientific research in order to monitor and evaluate the condition of the sea within the local maritime space and accepting that '*if the sea is sick, we'll feel it, if it dies, we die and that our future and the state of the ocean are one*' (Dr Sylvia Earle, 1995),

We, the people of Algoa Bay, commit ourselves to work with others to achieve the economic, social and environmental goals of ocean sustainability. We affirm our support of the Mission Blue goal to ignite public support for a global network of Hope Spots large enough to save and restore the blue heart of the planet. We accept that there is still time for us to take action to stabilize and reverse the global decline of the natural ocean systems that keep us alive. We acknowledge that although humans pose grave danger to the health of the ocean, we also represent great hope. As custodians of Algoa Bay we accept our responsibility to protect and sustain the blue heart of our maritime space.

Port Elizabeth, 11 December 2014

Conclusion

In many respects the Mission Blue Hope Spot initiative, implemented in South Africa by the SST, encourages local communities to become responsible stewards of their own resources in their own maritime domain. On the one hand humans are the main cause of the coastal and marine problems, but on the other hand humans can become solutions to the problems.

Yet this will require a massive coastal and ocean education and awareness program which Hope Spots are able to deliver, with support from the ocean business community and coastal authorities. If a world famous oceanographer such as Dr Sylvia Earle believes that there is still time to save the oceans, then there is hope, and coastal and ocean stakeholders should grasp this opportunity with both hands. Local people need to offer voluntary services wherever they can make a difference, and the ocean business community need to fund awareness programs. Like all successful programs, initiatives should start at home, and then consider how to spread local success to other areas of the globe.

> Many of us ask what can I do as one person but history shows us that everything good and bad starts because somebody does something or does not do something.
>
> Sylvia Earle, Oceanographer

References and further reading

Berkes, F. (2012). *Sacred Ecology*, 3rd edition. London and New York: Routledge.

Carson, R.L. (1951). *The Sea Around Us*, reissued 2003 with additions by J.S. Levinton and A.H. Zwinger. Oxford: Oxford University Press.

Dallmeyer, D.G. (2005). 'Toward a sea ethic' in: E.A. Norse and L.B. Crowder (eds), *Marine Conservation Biology: The science of maintaining the sea's biodiversity*. Washington, DC: Island Press, p.410–421.

Ehler, C. (2015). 'International progress in maritime spatial planning around the world'. Extended abstract of presentation at OECD workshop on 'The Future of Maritime Spatial Planning and Ocean Monitoring: What potential for economic tools and satellite technology?' Lisbon, June 2015.

European Commission (2011). 'Study on the economic effects of maritime spatial planning'. http://ec.europa.eu/maritimeaffairs/documentation/studies/documents/economic_effects_maritime_spatial_planning_en.pdf (accessed: May 2017).

Gray, N.J. and Campbell, L.M. (2008). 'Science, policy advocacy, and marine protected areas'. *Conservation Biology*, 23(2): 460–468.

Hale, L. (2009). 'Analyzing the role and feasibility of marine conservation agreements in protecting marine and coastal biodiversity in the Coral Triangle, Eastern Tropical Pacific, Gulf of California and Gulf of Mexico'. Proposal to the Walton Family Foundation. www.waltonfamilyfoundation.org (accessed: May 2017).

Halpern, B.S., Diamond, J., Gaines, S., Gelcich, S., et al. (2012). 'Near-term priorities for the science, policy and practice of coastal and marine spatial planning (CMSP)'. *Marine Policy*, 36(1): 198–205.

Huang, W., Corbett, J. and Jin, D. (2015). 'Regional economic and environmental analysis as a decision support for marine spatial planning in Xiamen'. *Marine Policy*, 51(Jan): 555–562.

IUCN (International Union for Conservation of Nature)/WCPA (World Commission on Protected Areas) (2008). 'Establishing marine protected area networks – Making it happen'. https://cmsdata.iucn.org/downloads/mpanetworksmakingithappen_en.pdf (accessed: May 2017).

Jay, S. and Gee, K. (2014). *TPEA Good Practice Guide: Lessons for cross-border MSP from transboundary planning in the European Atlantic*. Liverpool, UK: University of Liverpool.

Jay, S., Flannery, W., Vince, J., Liu, W-H., et al. (2012). 'Coastal and marine spatial planning: International progress in marine spatial planning'. *Ocean Yearbook*, 27(1): 171–212.

Kerwath, S.E., Winker, H., Gotz, A. and Attwood, C.G. (2013). 'Marine protected area improves yield without disadvantaging fishers'. *Nature Communications*, 4: 23–47.

Leopold, A. (1949). *A Sand County Almanac: And sketches here and there*. Oxford: Oxford University Press.

Morris, W.F. (2005). *The Bay of Lost Cargoes: Being a record of the shipwrecks of Algoa Bay*. Self-published.

The Nature Conservancy (2010). 'The Nature Conservancy annual report 2010'. www.nature.org/media/annualreport/annualreport2010.pdf (accessed: May 2017).

The Nature Conservancy and Conservation International (2012). *Practitioner's Field Guide for Marine Conservation Agreements: Best practices for integrating rights-based incentive agreements into ocean and coastal conservation efforts*. Narragansett, RI: The Nature Conservancy.

The Nature Conservation (TNC) (2010). *The Role and Feasibility of Marine Conservation Agreements in the Western Indian Ocean*. York, UK: University of York.

Olver, C. (2008). 'Report on issues relating to the potential relocation of the Port Elizabeth Manganese Terminal and Tank Farm to the Port of Ngqura'. www.8linkd.com/sites/default/files/files/Report-Issues%20Relating%20to%20the%20Potential%20Relocation%20of%20the%20Port%20Elizabeth%20Manganese%20Terminal%20and%20Tank%20Farm%20to%20the%20Port%20of%20Ngqura%20-%20Olver%202008.pdf (accessed: May 2017).

Organisation for Economic Co-operation and Development (OECD) (2016). 'The economics of marine protected areas: Work in progress'. www.oecd.org/environment/resources (accessed: April 2017).

Safina, C. (2003). 'Launching a sea ethic'. *Wild Earth*, 12(4): 2–9.

Spalding, M.D., Meliane, I., Milam, A., Fitzgerald, C., et al. (2013). 'Protecting marine spaces: Global targets and changing approaches'. *Ocean Yearbook*, 27: 213–249.

UNEP (United Nations Environment Programme) (2015). 'Regional state of the coast report Western Indian Ocean'. http://web.unep.org/nairobiconvention/regional-state-coast-report-western-indian-ocean-0 (accessed: April 2017).

WCPA (World Commission on Protected Areas)/IUCN (International Union for Conservation of Nature) (2007). 'Establishing networks of marine protected areas: A guide for developing national and regional capacity for building MPA networks'. Non-technical summary report. www.cbd.int/doc/pa/tools/Establishing%20Marine%20Protected%20Area%20Networks.pdf (accessed: May 2017).

World Wide Fund (WWF) (2014). 'State of management of South Africa's marine protected areas'. WWF-SA, Report Series. http://assets.wwfza.panda.org/downloads/mpastateofmanagementreport04nov2009weblowerdpi.pdf (accessed: April 2017).

Worm, B., Barbier, E.B., Beaumont, N., Duffy, J.E., et al. (2006). 'Impacts of biodiversity loss on ocean ecosystem services'. *Science*, 314: 787–790.

Chapter 4

Managing conflicts of interest

There are no 'one size fits all' actions that can deliver a blue economy. Some are global in scale (such as governance of the high seas), but many are needed at local, national and regional levels. Managing the multiple uses of the ocean is therefore vital. Successful management of any activity, however, requires adequate understanding both of the activity and of the context in which it takes place. Such an understanding is needed even more when the management tasks are divided among many players: unless each knows how the part they have to play fits into the overall pattern, there are risks of confusion, contradictory actions and failures to act. Managing human uses of the ocean has inevitably to be divided among many players, who may have conflicts of interest. This will be illustrated by three examples, from the Sargasso Sea, the Mediterranean and Hawaii.

The Sargasso Sea

Over 70% of the surface of planet Earth is covered by the oceans. About 64% of this watery coverage is classified as high seas. It is a deep and open ocean beyond the jurisdiction or control of any individual country, with its own range of habitats, species and unique ecosystems. As such there are few measures providing protection for the high seas in the same ways as those enjoyed by terrestrial and inshore environments. In recent years there has been growing awareness of and concern for the need to resolve this lack of protection; a need which grows more urgent as the open oceans and deep seas become increasingly exploited.

The Sargasso Sea is a fundamentally important part of the world's ocean, located within the North Atlantic sub-tropical gyre with its boundaries defined by the surrounding currents. It is the only sea without land boundaries with water depths ranging from the surface coral reefs of Bermuda to abyssal plains at 4,500 m. The Sargasso Sea's importance derives from the interdependent mix of its physical structure and properties, its ecosystems, its role in global scale ocean and earth system processes, its socio-economic and cultural values, and its role in global scientific research. Despite this, the Sargasso Sea is threatened by a range of human activities that either directly adversely impact it or have the potential to do so. Being open ocean, the Sargasso Sea is beyond the jurisdiction and responsibility of any national government, and as such it has little protection.

The Sargasso Sea Alliance

In 2009 Bermuda decided to improve the stewardship of their surrounding seas, both within their exclusive economic zone (EEZ) and into the wider high seas. With leading conservation and marine science organisations, they began to investigate opportunities within current mechanisms for high seas governance with the aim of affording protection for the Sargasso Sea. This resulted in the formation of the Sargasso Sea Alliance, a consortium of the Government of Bermuda, leading conservation and marine science organisations and individuals. The Alliance partners are:

- International Union for the Conservation of Nature (IUCN) and its World Commission on Protected Areas
- Mission Blue/Sylvia Earle Alliance
- Marine Conservation Institute
- Woods Hole Oceanographic Institution
- Atlantic Conservation Partnership
- Bermuda Institute for Ocean Sciences (BIOS)
- Bermuda Underwater Exploration Institute (BUEI)
- WWF International.

Formed to draw attention to the international importance of the Sargasso Sea and the threats upon it, the Sargasso Sea Alliance seeks to establish appropriate protective measures through the various sectoral organisations with interests in the Sargasso Sea, such as the International Commission for the Conservation of Atlantic Tuna (ICCAT: for fisheries), the International Maritime Organisation (IMO: for shipping), and the International Seabed Authority (ISA: for seabed exploitation issues). Creating a network of protection helps to mitigate the lack of overall protective measures due to high seas status, leading to the creation of an open ocean marine protected area (MPA).

The Sargasso Sea Alliance report (2009), entitled 'The protection and management of the Sargasso Sea', provides a summary of the scientific and other supporting evidence for the importance of the Sargasso Sea and is intended to: develop international recognition; start the process of establishing appropriate management and precautionary regimes within existing agreements; and stimulate a wider debate on appropriate management and protection for the high seas. The Sargasso Sea is a place of legend with a rich history of great importance to Bermuda. Other attributes of the Sargasso Sea, as described in the report, include:

- It is an iconic ecosystem based upon floating *Sargassum*, the world's only holopelagic seaweed, hosting a rich and diverse community including ten endemic species.

- It provides an essential habitat for nurturing a wide diversity of species, many of which are endangered or threatened.
- It is the only breeding location for the threatened European and American eels.
- It lies within a large ocean gyre which concentrates pollutants and which has a variety of oceanographic processes that impact its productivity and species diversity.
- It plays a disproportionately large role in global ocean processes of carbon sequestration.
- It is of major importance for global scientific research and monitoring and is home to the world's longest ocean time series of measurements.
- It has significant values to local and worldwide economies.
- It is threatened by activities including overfishing, pollution, shipping and *Sargassum* harvesting.

Apart from overfishing, many of the threats are potential, with few direct causal relationships between specific activities and adverse impacts. However, there is accumulative evidence that the Sargasso Sea is being adversely impacted by human activities, and with the possibility of new uses for *Sargassum* in the future, the lack of direct scientific evidence does not preclude international action through the established precautionary approach. The opportunity to recognise the importance of the Sargasso Sea and to develop and implement procedures to protect this iconic region and the wider high seas should be taken now.

There are four overarching aims of the Sargasso Sea Alliance:

- To build an international partnership to secure global recognition of the importance and ecological significance of the Sargasso Sea, the threats that it faces, and the precautionary management it needs.
- To use existing regional, sectoral and international organisations to secure a range of protective measures for the Sargasso Sea.
- To establish appropriate management for the Sargasso Sea.
- To use the current process as an example of what can and cannot be delivered through existing frameworks in marine areas beyond national jurisdictions to inform the global debate and to provide a model for protection of other high seas regions.

The success of the Alliance will be measured by international and public recognition of the importance of the Sargasso Sea, by approval and implementation of appropriate protection measures, and by the establishment of capacity to monitor their implementation. Underpinning these ambitions is the need for a robust science case, which, together with supporting evidence, demonstrates the importance and values of the Sargasso Sea and the reasons why improved protection and management are both necessary and urgent. This summary science and evidence case provides the information needed to begin the political process necessary to deliver the aspirations of the Alliance and the Government of Bermuda. To assist the preparation of this summary case a number of reports were specially commissioned by the Sargasso Sea Alliance, and are available for download at: www.sargassoalliance.org.

The Sargasso Sea Alliance initiative is already having an impact. It is informing the global ocean protection debate, and is bringing together existing knowledge of the Sargasso Sea so that its importance and value can be readily appreciated. Whether viewed environmentally, socially, economically or as a critical area for global marine research, the Sargasso Sea is hugely more valuable as an intact and healthy ocean area than as one that is depleted and degraded. The summary of this evidence-based case study provides information on the nature and importance of the Sargasso Sea as a necessary prerequisite for seeking international recognition and appropriate agreements to better manage human impacts upon it.

The Sargasso Sea Alliance is an excellent example of collaboration in the high seas. Maritime clusters at the coast and alliances in the high seas both incorporate the essential ingredients of cooperation in a multiuser environment where a single structure would not be able to accommodate so many partners with diverse interests. The forging of multiuser cooperative alliances is mutually beneficial and encourages a collective approach in which responsible ocean governance can flourish.

The Barcelona Convention

In 1975, 16 Mediterranean countries and the European Community adopted the Mediterranean Action Plan (MAP), the first-ever Regional Seas Programme under the United Nations Environment Programme (UNEP)'s umbrella. In 1995, the Action Plan for the Protection of the Marine Environment and the Sustainable Development of the Coastal Areas of the Mediterranean (MAP Phase II) was adopted by the contracting parties to replace the Mediterranean Action Plan of 1975. Today, the Barcelona Convention and MAP are more active than ever. The contracting parties now total 22, and they are determined to protect the Mediterranean marine and coastal environment while boosting regional and national plans to achieve sustainable development.

The contracting parties are: Albania, Algeria, Bosnia and Herzegovina, Croatia, Cyprus, Egypt, the European Community, France, Greece, Israel, Italy, Lebanon, Libya, Malta, Monaco, Montenegro, Morocco, Slovenia, Spain, Syria, Tunisia, and Turkey. The Convention's main objectives are:

- To assess and control marine pollution.
- To ensure sustainable management of natural marine and coastal resources.
- To integrate the environment in social and economic development.

- To protect the marine environment and coastal zones through prevention and reduction of pollution, and as far as possible, elimination of pollution, whether land- or sea-based.
- To protect the natural and cultural heritage.
- To strengthen solidarity among Mediterranean coastal states.
- To contribute to improvement of the quality of life.

Protocols

The Barcelona Convention has given rise to seven protocols addressing specific aspects of Mediterranean environmental conservation:

- Dumping Protocol (from ships and aircraft)
- Prevention and Emergency Protocol (pollution from ships and emergency situations)
- Land-based Sources and Activities Protocol
- Specially Protected Areas and Biological Diversity Protocol
- Offshore Protocol (pollution from exploration and exploitation)
- Hazardous Wastes Protocol
- Protocol on Integrated Coastal Zone Management (ICZM).

Structure

The contracting parties decide on MAP strategies, budget and programme in pursuit of MAP's goal at their ministerial-level meetings, held every two years. They appoint Focal Points to review the progress of work and ensure the implementation of recommendations at the national level. A rotating bureau of six representatives of the contracting parties guides and advises the MAP Secretariat (located in Athens) in the interim period between the biannual meetings. MAP's activities are primarily financed by the contracting parties through their contributions to the Mediterranean Trust Fund. Other main sources of funding to support specific projects and activities include voluntary contributions from the European Union, UN agencies, and the Global Environment Facility (GEF).

Key priorities

Key MAP priorities for the coming decade are:

- To bring about a massive reduction in pollution from land-based sources.
- To protect marine and coastal habitats and threatened species.
- To make maritime activities safer and more conscious of the Mediterranean marine environment.
- To intensify integrated planning of coastal areas.

- To monitor the spreading of invasive species.
- To limit and intervene promptly on oil pollution.
- To further promote sustainable development in the Mediterranean region.

Horizon 2020

Euro-Mediterranean governments aim to tackle the top sources of Mediterranean pollution by the year 2020 through the Horizon 2020 initiative that is built around 4 elements:

- Projects to reduce the most significant pollution sources focussing on industrial emissions, municipal waste and urban wastewater, responsible for up to 80% of pollution in the Mediterranean Sea.
- Capacity-building measures to help neighbouring countries create national environmental administrations that are able to develop and police environmental laws.
- Using the Commission's Research budget to develop and share knowledge of environmental issues relevant to the Mediterranean.
- Developing indicators to monitor the success of Horizon 2020.

The initiative is open to all stakeholders and calls for the mobilisation of the necessary financial resources and the provision of technical support from all partners. All stakeholders are invited to work together to maximise assistance, exploit opportunities and meet the commitments of the timetable.

Projects

Prevention of and response to environmental damage from offshore exploration and exploitation activities are important issues for the EU. On 17 December 2012 the Council approved EU accession to the Offshore Protocol, thus underlining EU commitment to reducing environmental impacts of offshore activities in the Mediterranean through efficient regional cooperation. The legal consequence of this Council act is that the Offshore Protocol is now a part of EU legislation. On 10 June 2013 the EU adopted a Directive on safety of offshore oil and gas operations.

A project managed by the UNEP/MAP Coordinating Unit and co-financed by the EU contributed substantially to the implementation of the Ecosystem Approach that the contracting parties of the Barcelona Convention (and among them the EU) had agreed to implement in the Mediterranean. The Ecosystem Approach is compatible with EU environment policies, in particular the EU Marine Strategy Framework Directive. The implementation of EU and broader Mediterranean policies for conservation and, where necessary, restoration, of the marine and coastal environment, is mutually supportive.

On 25 October 2011, the European Commission adopted a proposal for the EU to accede to a Protocol of the Barcelona Convention that protects the Mediterranean against pollution from offshore exploration and exploitation activities. Such activities are increasing in the Mediterranean, and the area is particularly vulnerable due to its semi-closed configuration and significant seismic activity. The Offshore Protocol covers a wide range of exploration and exploitation activities and touches upon permit requirements, the removal of abandoned or disused installations, the use and removal of harmful substances, and provisions on safety, contingency planning and monitoring. On 17 December 2012 the Council approved EU accession to the Offshore Protocol, thus underlining EU commitment to reducing environmental impacts of offshore activities in the Mediterranean through efficient regional cooperation. The legal consequence of this Council Act is that the Offshore Protocol now becomes part of EU legislation.

Hawaii: Recreation carrying capacity and management research study

In 2008 a research study was undertaken by a project team of scientists and academics from Oregon State University, Department of Forest Ecosystems and Society. The project was funded by the Hawaii Coral Reef Initiative.

Background

Hawaii hosts approximately seven million visitors each year who spend more than US$11 billion. In the last 20 years tourism has increased by over 65%. More than 80% of Hawaii's visitors engage in recreation activities in the state's coastal and marine areas with the majority of these individuals participating in snorkelling or diving. Other popular coastal recreation activities include ocean kayaking, swimming, sunbathing, beach walking, and surfing. Beaches and reefs are also important resources for local residents: for example, approximately 30% of households in Hawaii have at least one person who fishes for recreation.

As popularity of Hawaii's coastal areas continues to increase, demand for access and use can disrupt coastal processes, damage ecological integrity of sensitive environments, reduce the quality of user experiences, and generate conflict among stakeholders regarding appropriate management responses. As a result, agencies are faced with challenges that include estimating use thresholds (i.e. carrying capacities) and how to manage and monitor use levels to ensure that thresholds are not violated and user experiences are not compromised by such issues as crowding and conflict. The purpose of the research study project, therefore, was to examine carrying capacity, conflict and management, as related to recreation use at Kailua Beach Park on the east coast of Oahu, Hawaii.

Objectives

The project aims to:

- Use social science approaches to measure, determine and inform social and facility indicators of recreation carrying capacities, and to determine thresholds when perceived impacts for these indicators reach unacceptable levels.
- Estimate the extent to which indicators of recreation carrying capacities are currently being exceeded and if this is impacting user experiences.
- Measure support and opposition toward management strategies for minimizing coastal recreation impacts (e.g. education and limitation of use), and how situational factors (e.g. reef damage andr litter) differentially influence support of these strategies.
- Determine the extent to which user conflicts exist both within and among various recreation activity groups.
- Compare the extent to which evaluations of coastal recreation impacts differ among groups (e.g. visitors versus locals and various tourism/recreation activity groups).

Other concepts to be examined in this project include recreationists' satisfaction with current conditions, perceptions of crowding, value orientations toward coastal environments, likelihood of future use, changes in use patterns (e.g. temporal and spatial displacement), and socio-demographic characteristics.

CASE STUDY: HAWAII, HANAUMA BAY

Hanauma Bay Nature Preserve is one of the most popular recreational swimming, snorkelling, diving and picnicking areas in Hawaii. The bay is legally designated as a State Underwater Park and Conservation District. There are several accounts of how Hanauma Bay got its name: *hana* means 'bay' and *uma* has several meanings, one being 'curved'; therefore one translation of Hanauma is 'curved bay', which is certainly appropriate. (The name is pronounced *ha-now-ma*, with the accent on 'now'.)

In 1950, the City and County of Honolulu built a road to the beach at Hanauma Bay and installed new restrooms and showers, helping it to quickly become a favourite fishing and picnicking park for residents of Honolulu. Swimming at Hanauma, however, was difficult: except for the two natural lagoons at the north end of the beach, an almost unbroken expanse of shallow reef fringed the shore. A photo from this period shows a narrow channel running through the reef, possibly dug at this time to allow swimmers to reach deeper water.

In 1956, a 200 ft. wide swath was blasted through the reef. Tons of coral rock were piled on the beach and removed. In a matter of days, the central portion of Hanauma's beautiful fringing reef, and some of the reef beyond, had been destroyed forever. Such devastation would be unthinkable today, but in those times, it was considered as improvement (i.e. the public now had more swimming space). On the positive side, the dynamiting and dredging created a new biological zone inside the reef. Loose rubble left by the blasting became habitat for a host of small invertebrates and the fishes that feed on them, many of which had probably not lived inside the reef in any numbers before. The swimming area (knee- to chest-deep at low tide) gave these fishes ample living space. The channel provided a way in and out, as well as improving water circulation inside the reef. There are quite a few species occurring today in the artificial swimming area that do not occur in the natural sandy-bottomed lagoons.

With more places to swim and easier access to deep water, Hanauma Bay increased in popularity. Many residents fished there and, by the 1960s, few fish of any size were left. In 1967, and not without opposition, the State Division of Fish and Game declared the entire bay a Marine Life Conservation District, prohibiting the taking of marine life, shells, coral, rocks or sand. However, one problem remained: surf entering through the wide cable channel was eroding the beach.

In 1970, the City Department of Parks and Recreation remedied this by partially blocking the channel with an artificial reef of heavy basalt boulders. As part of the same project, and in spite of some community opposition, the City blasted and dredged more reef on the right side of the bay to create a second swimming area in front of the far restroom. Workers buried the dredgings under the beach and covered them with 4,000 cubic ft. of white sand brought in from O'ahu's north shore. As before, the blasting and dredging created a rubble bottom in which many small animals could live and increased the habitat for larger fishes inside the reef. Also, the new boulder pile sheltered the original swimming area from incoming waves, making it more habitable for marine life. Its outer face became one of the best shallow snorkelling areas outside the reef.

After creation of the marine preserve, snorkelling and fish feeding at Hanauma became hugely popular. Tourists brought in by the busload soon began to outnumber residents on the beach. To make room for everyone, the City built an additional parking lot and by the late 1980s as many as 10,000 people per day visited the bay. The number of fish in the natural lagoons and artificial swimming areas increased enormously as people fed them bread, peas and human snack foods. Tossing a handful of fish food into the water caused it to boil with fishes large and small. It was thrilling to stand waist deep with fast-moving animals swirling around: most visitors had never been so close to wild creatures. Fish feeding, of course, was not the only problem. The beach and near-shore waters were overcrowded. Visitors left their vehicles on the lawn, on the roadways and on the highway. Commercial tours were monopolizing the picnic facilities. Also, many people believed at the time that beachgoers were killing the coral by walking on it. Although unsubstantiated, this environmental concern prompted action.

Around 1990 park authorities began a program to minimize impact on the bay both by reducing the number of visitors and by teaching conservation. Most commercial activities were banned and stricter controls on parking were put in place. When the lot filled, cars were turned away. Meanwhile, a volunteer organization, the Friends of Hanauma Bay, working from a booth on the beach, taught thousands of visitors to care for and appreciate marine life. In 1998 the City began charging admission and parking fees, further reducing the number of visitors. The program was a success, albeit with an unforeseen consequence: local people came in fewer numbers and Hanauma Bay became largely a tourist destination. Park officials hope to reverse this. In 1999 the City began a three-year carrying capacity study to determine the number of visitors that is environmentally safe.

Happily, results released in June 2000 suggested that beachgoers and snorkelers have had little detrimental effect on marine life. Besides fish feeding, which has ceased, the study identified only a few minor negative impacts: runoff from freshwater showers and swimmers not using restrooms. It was too soon to determine whether people walking on the reef over the years had significantly altered or damaged it. About 3,000 people now visit Hanauma Bay Nature Preserve each day, making it one of the most visited beaches in Hawaii. To keep the marine environment as natural as possible without further lowering the visitor count, the City and County of Honolulu has moved the snack bar up from the beach and built a new Education Centre (see Figure 4.1, which also outlines other proposed changes to the area). There is now a mandatory training video for first-time visitors: all beachgoers hear a clear message of preservation, conservation and safety before entering the water. Today the City and County of Honolulu administers the part of Hanauma Bay Nature Preserve lying above the high tide line, and the State of Hawaii Department of Land and Natural Resources manages the underwater portion.

Conclusion

The Hanauma Bay case study illustrates both the positive and negative impact of tourism. By building an easy access road from Honolulu to

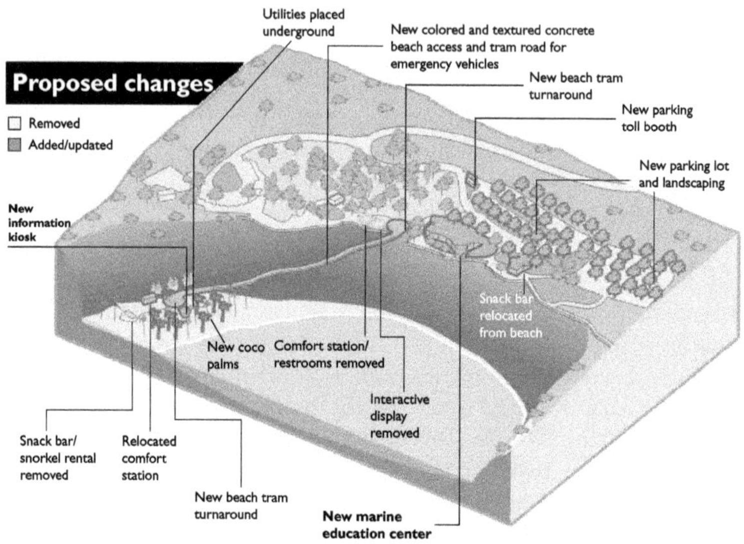

Figure 4.1 Proposed changes at Hanauma Bay, Hawaii

Hanauma the authorities increased the number of visitors significantly to the extent that overcrowding over a period of time was causing ecological and environmental degradation. Wisely they calculated the carrying capacity and determined a threshold to control the crowds.

On the positive side, by introducing a compulsory education and awareness program the authorities were able to change the attitude and behaviour of visitors once they were more aware of their responsibilities. Once again we can see the benefits of education and awareness as the most successful tool for developing responsible recreational users. What is encouraging is that this coastal gem has recovered. The 3,000 daily users value the coastal tourism experience, reminding us of what Jacques Cousteau said many years ago, 'People will protect what they love (enjoy)'.

Note: This case study was developed by the author from information posted on www.hanaumabaystatepark.com and *Hanauma Bay Education Program*, Ka Pili Kai, published quarterly by the University of Hawai'i Sea Grant College Program (UH Sea Grant), School of Ocean and Earth Science and Technology (SOEST).

Discussion

The Sargasso Sea Alliance illustrates that for the high seas the most appropriate form of collaboration is the forging of mutually beneficial cooperative alliances, i.e. collective ocean governance. The management research study in Hawaii illustrates the value of beach surveys to discover what recreational users like and dislike. Surveys are a useful management tool to ensure that thresholds are not violated and user experiences are not compromised by such things as crowding and conflict. The Barcelona Convention illustrates how 22 contracting parties can collaborate to protect the marine and coastal environment while boosting regional and national plans to achieve sustainable development. This collaboration has resulted in many Mediterranean resorts reinventing themselves to woo back tourists who were visiting other destinations that were cleaner, greener and less crowded.

References and further reading

Barker, N.H.L. and Roberts, C.M. (2004). 'Scuba diver behaviour and the management of diving impacts on coral reefs'. *Biological Conservation*, 120: 481–489.

Cesar, H.S.J., van Beukering, P., Dierking, J., Pintz, S., et al. (2004). *Economics of MMAs in Hawaii: Interim results*. Honolulu: Social Science Research Institute, University of Hawaii.

CSV Consultants (2007). 'Kailua ocean recreation study'. Unpublished document: Honolulu, Hawaii.

Davidson, K., Hamnett, M. and Minato, C. (2003). *Economic Value of Hawaii's Nearshore Reefs*. Honolulu: Social Science Research Institute, University of Hawaii.

Dinsdale, E.A. and Fenton, D.M. (2006). 'Assessing coral reef condition: Eliciting community meanings'. *Society and Natural Resources*, 19: 239–258.

Jones, C. (n.d.). 'Dr Beach names Hawaii's Hanauma Bay best beach in America'. *The Morning Journal*. www.morningjournal.com/article/MJ/20160526/ENTERTAINMENT/160529630 (accessed: April 2017).

Kerr, W., Bos, M. and Clark, A. (2005). *Hawaii's Local Action Strategy (LAS) to Address Recreational Impacts to Reefs*. Honolulu: Department of Land and Natural Resources, Hawaii Ecotourism Association.

Lankford, S.V., Inui, Y., Whittle, A., Luna, R., et al. (2005). 'Sustainability of coastal marine recreation: Modeling social carrying capacity for Hanauma Bay, Hawaii'. Final project report for the University of Hawaii Sea Grant College Program SOEST. Cedar Falls, IA: University of Northern Iowa.

Lynch, T.P., Wilkinson, E., Melling, L., Hamilton, R., et al. (2004). 'Conflict and impacts of divers and anglers in a marine park'. *Environmental Management*, 33: 196–211.

National Oceanic and Atmospheric Administration (NOAA) (2009). 'Coral reef habitat assessment for U.S. marine protected areas: State of Hawaii: NW and Main Hawaiian Islands'. www.coris.noaa.gov/activities/habitat_assessment/hawaii.pdf (accessed: April 2017).

Needham, M.D., Tynon, J.F., Ceurvorst, R.L., Collins, R.L., et al. (2008). 'Recreation carrying capacity and management at Kailua Beach Park on Oahu, Hawaii'. Final project report for the Hawaii Coral Reef Initiative – Research Program. Corvallis: Department of Forest Ecosystems and Society, Oregon State University.

QMark Research and Polling (2005). 'Noneconomic value of Hawaii's coral reef: Results from telephone survey and focus groups'. Honolulu: Social Science Research Institute, University of Hawaii.

Rodgers, K.S. and Cox, E.F. (2003). 'The effects of trampling on Hawaiian corals along a gradient of human use'. *Biological Conservation*, 112: 383–389.

Sargasso Sea Alliance (2009). 'The protection and management of the Sargasso Sea'. www.sargassoseacommission.org/storage/documents/Sargasso.Report.9.12.pdf (accessed: April 2017).

Tissot, B.N. and Hallacher, L.E. (2000). 'Diver impacts on coral reefs at Kealakekua Bay, Hawai'i'. www.coralreefnetwork.com/research/kbay (accessed: April 2017).

van Beukering, P.J.H. and Cesar, H.S.J. (2004). 'Ecological economic modeling of coral reefs: Evaluating tourist overuse at Hanauma bay and algae blooms at the Kihei coast, Hawai'i'. *Pacific Science*, 58: 243–260.

Chapter 5

Marine and coastal aquaculture

Aquaculture, tourism and recreation all require similar standards of water and environmental quality and good local infrastructure, so there is potential for conflicts of interest and competition to arise. Aquaculture is a very important industry for many countries, especially where communities depend on the employment and revenue it provides. The location of fish farms could increase the risk of human predator interaction, especially near beaches. Marine aquaculture or 'mariculture' (for a review, see Stead 2015) gives rise to a number of concerns, for example the placement of fish farms within sensitive marine areas, the spread of invasive species, and dependency on industrial catches of wild fish to feed fish in mariculture. Where the demand for land for agriculture or aquaculture (e.g. to increase production of rice and fish for local consumption) or for infrastructure development necessitates the conversion of mangrove areas, the decision should be based on the results of a thorough environmental impact assessment, including a valuation of all the direct and indirect benefits mangroves provide. Proper measures should be taken to minimize the damage to the mangrove ecosystem as a whole. This means the establishment of a strategic planning mechanism and institutional framework involving local, provincial and national authorities in forestry, agriculture, fisheries and rural development, as well as other stakeholders (e.g. communities, NGOs, private sector, and funding and technical agencies) in the affected areas to provide advice on land-use options within an integrated coastal area management framework. In this regard maritime clusters could make a valuable contribution by encouraging a high level of multi-stakeholder collaboration.

Introduction

Since time immemorial, the coastal zone has been a center of human activity as a result of its high biological productivity and easy accessibility. Close to half of the world's population resides within 100 km of a coastline. The wide variety of goods and services provided by mangroves, seagrasses, coral reefs and other coastal ecosystems include the production of aquatic plants and animals used for food, medicines, construction and other human needs (e.g. recycling of nutrients and filtration of pollutants; control of flooding and soil erosion; and protection from typhoons, storm surges and tsunamis).

The global value of coastal ecosystems has been estimated at US$12.57 trillion/year and those of mangrove forests, together with tidal marshes, at US$1.65 trillion/year (FAO 2014). The many uses of the coastal zone include artisanal and commercial fisheries; aquaculture; agriculture; human settlements; harbors, ports and navigation; recreation and tourism; and mining and industries. Such multiple uses have given rise to conflicts over resource use. In recent years, some of the most controversial conflicts have been related to the apparent and potential negative impacts of aquaculture.

In 2002, aquaculture worldwide produced a total of 51.4 million metric tons (mmt), valued at US$60 billion (FAO 2014). Freshwater fish contributed 47.7% of volume, followed by mollusks (22.9%), plants (22.6%) and crustaceans (4.2%). Aquaculture harvests were predominantly from Asia (91.2%), followed by Europe (4.0%) and the Americas (3.6%). Around 55% of production in 2001 came from land-based ponds and water-based pens, cages, long-lines and stakes in brackish water, and marine habitats; freshwater habitats contributed 45%. Mangrove conversion to shrimp ponds is the single major factor that has contributed to the negative press received by aquaculture.

Widely projected to compensate for declining fisheries catches, aquaculture now provides a third of total fisheries production compared to only 16% in 1991. However, the phenomenal growth of aquaculture in recent years, combined with a tendency to ease out other stakeholders and convert multiuse coastlines to a single-use resource, have given rise to social disruption and conflicts. A typical example is Dagupan in the Philippines where mangroves, tidal creeks, estuaries and near shore areas traditionally accessible to small-scale fishers, intertidal gleaners and other local folks have been converted first from mangroves to fishponds in the 1950–1980s, then later to pens and cages both in estuarine waterways and sub-tidal habitats such as seagrass beds.

Mangroves and coastal aquaculture

To what extent are mangroves and aquaculture competing land uses? How should different land-use options be reconciled, and what should be done to ensure that future developments in coastal aquaculture become sustainable land-use options? Many types of coastal aquaculture are practiced within or close to mangrove areas. Many of these have no or very limited adverse effects on mangroves and have been practiced for hundreds of years providing livelihoods to thousands of rural families. Although there may be issues of localized water pollution/eutrophication arising from intensive cage culture of fish, the main concern for mangrove areas and their associated creek systems is related to the construction of shrimp and fish ponds, particularly where these involve the clearing of existing mangroves.

Any decision to convert a mangrove area for aquaculture should be based on the results of a thorough environmental impact assessment. A strategic planning mechanism should be established to minimize the damage to the mangrove ecosystem as a whole. Local, provincial and national authorities in forestry, agriculture, fisheries and rural development, as well as other stakeholders, should provide advice on land-use options within an integrated coastal area management framework.

Environmental impacts of aquaculture

The coastal zone bears most of the ecological consequences of aquaculture development. These include habitat loss/modification, excessive harvesting of wild seed/spawners and damage to bycatch, introductions of exotic species, escapes of cultured animals, spread of diseases, interactions with wild populations, misuse of chemicals and antibiotics, release of wastes, and dependence on wild fisheries.

Habitat loss/modification

Mangrove conversion to shrimp ponds is the single major factor that has contributed to the negative press received by aquaculture. Southeast Asia has 35% of the world's 18 million ha of mangrove forests, but has also suffered from the highest rates of mangrove loss, (e.g. 70–80% in the Philippines and Vietnam for the last 30 years). Around half of the 279,000 ha of Philippine mangroves lost from 1951 to 1988 were developed into culture ponds; 95% of Philippine brackish water ponds from 1952 to 1987 were derived from mangroves.

Globally, more than a third of mangrove forests have disappeared in the last two decades, and shrimp culture is the major human activity accounting for 35% of such decline. This transformation results in loss of essential ecosystem services generated by mangroves, including the provision of fish/crustacean nurseries, wildlife habitat, coastal protection, flood control, sediment trapping and water treatment. Many aquaculture farms in Asia stock wild-caught juveniles rather than hatchery-reared post-larvae derived from wild spawners or broodstock. Collection of such 'seedstock' can have major consequences for wild fisheries in terms of high rates of bycatch. For example, juvenile milkfish collected for grow-out constitute only 15% of total finfish fry from inshore seine net catches; the remaining 85% are discarded and left to die on the beach.

Introductions of species/spread of diseases

The greatest number of introductions of exotic fish and crustacean species outside their natural range has occurred for aquaculture purposes. The potential negative effects of such introductions include the degradation of host environment, disruption of the host community, genetic degradation of the host stock, and the introduction of diseases and parasites. Disease risks are particularly acute: almost all of the 20 identified viruses in marine shrimp have been described in cultured animals. Impacts from disease have also been observed in the case of transfers of species within their range. In recent years, viruses, notably the Whitespot Syndrome Virus (WSSV) and Yellowhead Virus, have caused catastrophic multimillion dollar crop losses in shrimp farms across Asia. The origin of the WSSV pandemic has been traced to the import into Japan of infected hatchery produced *P. japonicus* from Chinese hatcheries in 1993. Since then, it has spread to China, Taipei, Korea, India, the Philippines and even tropical America.

The future for aquaculture

For aquaculture to fulfill the promises of food security and poverty alleviation without causing negative environmental and socio-economic effects, a more holistic approach is required. This must involve other stakeholders in order to avoid a sectoral focus on aquafarms. The key issues to consider and address are outlined below.

Integrated coastal zone management

The groups vulnerable to the negative effects of shrimp culture generally do not participate in the formulation and implementation of public policies, for example determining location of ponds, pens and cages; regulating farm activities; and environmental impact assessment preparation. Community participation in coastal zone management is essential if questions of social equity are to be satisfactorily addressed. In community-based coastal resource management, fisherfolk and other local residents are the *de facto* day-to-day managers of resources. Coastal zones should be delineated for fisheries, aquaculture, tourism and other uses through the process of integrated coastal zone management (ICZM). Allocation of activities to locations should be based on the carrying or assimilative capacity of the environment for a given use, protection of community resources, rehabilitation of degraded habitats, stakeholder needs, and mechanisms for conflict resolution. ICZM is based on the concept of the ecological footprint which incorporates not only inputs such as feed and seed, but also outputs such as effluent treatment. Grow-out systems need to reduce and mitigate negative impacts and improve sustainability through the development and wider dissemination of available on-farm technologies (e.g. recycling), and integration of aquatic species (e.g. polyculture of tilapia/fish and shrimp in ponds).

Mangrove-friendly aquaculture

Mangrove research has increased exponentially in the last 50 years. The total number of publications on mangroves indexed by the Web of Science exceeded 8,000 in 2010, and since 2006 has consistently surpassed those on salt marshes. Early mangrove research focused on basic issues such as floristics and faunistics but these are increasingly replaced by ecological assessment of function and evaluation of the capacity of mangroves for ecosystem services such as fisheries, shoreline protection, carbon export/sequestration and bioremediation of wastes.

Parallel to this shift in research focus is the progressive widening of the spatial scale of research, from earlier dominance of localized, small-scale (e.g. forest- or tree-level) matters to regional and global perspectives. Over the past five decades, discussion of mangrove ecosystems and management has focused on: (1) the dynamics of carbon fixation, storage and mineralization; (2) their nursery function; (3) shoreline protection; and (4) their land-building capacity. Mangrove management worldwide has been guided by the scientific paradigms in these areas.

In contrast to their relatively simple forest structure and low diversity, tropical mangroves are ranked amongst the most productive natural ecosystems globally, notwithstanding the variable methodologies applied (Alongi 2009). Loss of mangrove production to herbivory is variable (Sousa and Dangremond 2011), but generally only amounts to 2–3% of the overall budget. The bulk of mangrove is therefore processed through the detritus food chain. The contribution of mangrove detritus to faunal biomass is not ubiquitously significant (Bouillon et al. 2000; 2002; 2004; 2008) and may deviate from simple availability (Bouillon, Koedam, Raman and Dehairs 2002). Also, rapid mineralization can occur in the water column of the tidal channel (Kristensen, Bouillon, Dittmar and Marchand 2008).

Mangroves and aquaculture are not necessarily incompatible. For example, seaweeds, bivalves and fish (e.g. seabass or grouper) in cages can be grown in mangrove waterways. Such mangrove-friendly aquaculture technologies are amenable to small-scale, family-based operations and can be adopted in mangrove conservation and restoration sites. Brackish water culture ponds may not necessarily preclude the presence of mangroves. Dikes and tidal flats fronting early Indonesian Tambak were planted with mangroves to provide firewood, fertilizers and protection from wave action. Present-day versions of integrated forestry–fisheries–aquaculture can be found in the traditional gei wai ponds in Hong Kong, mangrove-shrimp ponds in Vietnam, aquasilviculture in the Philippines, and silvofisheries in Indonesia. Alternatively, mangroves adjacent to intensive ponds can be used to process nutrients in pond effluents.

Fish farming: The new driver of the blue economy

Millions of people around the world find a source of income and livelihood in the fisheries sector. The most recent estimates indicate that in 2010 there were 54.8 million people engaged in the primary sector of capture fisheries and aquaculture. Of these, an estimated 7 million people were occasional fishers and fish farmers (of whom 2.5 million in India, 1.4 million in China, 0.9 million in Myanmar, and 0.4 million each in Bangladesh and Indonesia). More than 87% of all people employed in the fisheries sector in 2010 were in Asia, followed by Africa (more than 7%), and Latin America and the Caribbean (3.6%). Approximately 16.6 million (about 30% of all people employed in the fisheries sector) were engaged in fish farming, and they were even more concentrated in Asia (97%), followed by Latin America and the Caribbean (1.5%), and Africa (about 1%) (FAO 2014).

The less developed countries (LDCs), mostly in sub-Saharan Africa and in Asia, and home to 20% of the world's population (1.4 billion people), remain very small in terms of their share of world aquaculture production (4.1% by quantity and 3.6% by value). Aquaculture, otherwise known as fish farming, is the controlled cultivation of freshwater and saltwater animals or plants. The aquaculture sector overall is highly diverse and fragmented, ranging from smallholder ponds in Africa providing a few kilos of fish per year to international companies with annual turnover in excess of US$1 billion. While capture fisheries production remains stable, aquaculture production keeps on expanding and is set to remain one of the fastest-growing animal food-producing sectors and, in the next decade, total production from both capture and aquaculture will exceed that of beef, pork or poultry. Much of this development has occurred in Asia, which also has the greatest variety of cultured species and systems. Asia is also perceived as the 'home' of aquaculture, as aquaculture has a long history in several areas of the region and knowledge of traditional systems is most widespread. Its systems can range from an intensive indoor system monitored with high-tech equipment through to the simplest production systems such as small family ponds in tropical countries. At the other end of the scale are high-tech systems, such as the intensive indoor closed units used in North America for the rearing of striped bass or the sea cages used in Chile and Europe for growing salmon and bream. With capture fisheries becoming increasingly unsustainable due to overfishing, aquaculture is expected to overtake capture fisheries in supplying the world's protein requirements in the future (FAO 2012). In the recent past, some developing countries in Asia and the Pacific (Myanmar and Papua New Guinea), sub-Saharan Africa (Nigeria, Uganda, Kenya, Zambia and Ghana) and South America (Ecuador, Peru and Brazil) have made rapid progress to become significant or major aquaculture producers in their regions.

Aquatic products are increasingly traded globally, the volume having increased significantly over the past 10 years. New markets have emerged, and new products have appeared in the market. With restrictions on fishing in certain seas, some aquaculture products found strong niche markets and became important commodities in aquatic food trade. Traceability and improved and value-added products entered into the market. Although it fluctuates, all in all, the price of cultured fish has declined over the past 10 years, making fish an affordable food commodity to many.

However, the vital contributions from fisheries and aquaculture to global food security and economic growth remain constrained by an array of problems. These include poor governance, weak fisheries management regimes, conflicts over the use of natural resources, the persistent use of poor fishery and aquaculture practices, a failure to incorporate the priorities and rights of small-scale fishing communities, and injustices relating to gender discrimination and child labour.

Over and above, the impacts of climate change are also posing threats to sustainable aquaculture development, thus requiring focused implementation of mitigation and adaptation strategies.

Interaction between fish farming and tourism

There are a number of studies suggesting that aquaculture and tourism may compete for some of the same resources. Fish farming and tourism both place certain demands on the coastal environment with a range of potential interactions. A report published by Royal Haskoning in Scotland (Nimmo, Cappell and Grant 2009) investigated the following:

- Competition for space (land and sea)
- Navigational conflicts between fish farms and tourism
- Visual impact of fish farms on tourism
- Tourist perception of the environmental impact of fish farms
- Aquaculture as a tourist attraction
- Tourism as a consumer of aquaculture products
- Conflict between fish farms and other recreational groups.

The visual impact of fish farms is considered a key topic. Many coastal zone activities relating to public or national resources are regulated and may require a license. The extent of licensing required in different countries will affect how easy it is for an industry to develop and become profitable. Regulation may therefore influence the extent to which aquaculture and tourism can effectively coexist and develop in an area.

In some countries, user conflicts are a result of poorly defined rights over resources, where it may be unclear who owns the space and what uses they are entitled to exercise. As a result of this it can be difficult for individuals to enforce and protect their rights. To reduce the conflict between aquaculture and tourism groups in an area it is important to develop a clear policy for the rights of use across the competing users (Alongi 2002). Evidence shows that conflicts have arisen in many countries and that there is often no straightforward solution to resolve the matter (Dempster and Sanchez-Jerez 2008). However, careful planning and control strategies, such as conflict resolution management, can be implemented to prevent conflicts from arising and minimize them if they occur. Site selection is a key factor in any aquaculture operation affecting success, sustainability and environmental effects.

Integrated planning

Like elsewhere in the world, tourism has been found to contribute substantially to the economy in Mediterranean countries. Marine aquaculture is being practiced in coastal Mediterranean waters but has difficulty in finding locations to compete successfully or to coexist with tourism. The exploitation of the coastal land for tourism development has resulted in high land prices which have become prohibiting for other forms of development, including aquaculture, which is considered a relatively new economic activity in the area (Stephanou 1998). There is often a lack of political support for aquaculture in the Mediterranean and so it is not promoted in terms of national or regional policies. Where legal framework does exist for ICZM, aquaculture is rarely supported as a positive element of local development, despite creating permanent employment, generating fresh produce and being a potential source of tourism activity (de Monbrison 2004).

The local authority of the Murcia region of Spain made an effort to promote regional planning for aquaculture development. All relevant sectors, including tourism groups, were invited to map areas that each sector considered to be suitable for aquaculture development. However there was no correlation between the maps produced by different groups, which led to the termination of the project for two years, after which it was decided that the fish farms would be located offshore (de Monbrison 2004).

A more positive outcome resulted from a study of tourism and aquaculture in Tenerife (Dempster and Sanchez-Jerez 2008) which recognized the clear dominance of tourism over other sectors. Of the coastline of Tenerife, 56% was identified as being suitable, with 46% being very suitable, suggesting that marine cage aquaculture could be developed in coexistence with the established tourism industry. Knowledge of the potential future developments on the island was taken into account so that aquaculture could be developed without directly competing for the same coastal space and resources as the tourism industry. Once the potential for developing marine aquaculture in Tenerife had been determined, further studies were then recommended to investigate site selection in terms of water quality requirements (such as temperature, suspended solids and sewage discharges), marine environment (such as waves, currents and bathymetry), infrastructure to support the industry (such as roads, airports and ports), and other uses of the coastal space, such as fisheries (Pérez, Telfer, Beveridge and Ross 2002; Pérez, Telfer and Ross 2005).

This approach primarily assesses compatibility with tourism before considering the location based on physical and environmental conditions required for aquaculture production. The approach is appropriate given the clear dominance of tourism to Tenerife over a relatively undeveloped aquaculture sector. For Scotland no such forward planning has been evident, creating a complex mosaic of coastal users that are already present and making demands for space.

The provincial government in British Columbia, Canada, undertook a series of coastal resources interest studies to identify areas along the coast with high or low potential for salmon farming, along with possible resource conflicts. User groups and resource agencies mapped the major uses along the coast. Draft maps of finfish aquaculture opportunities were then prepared and presented at public open houses, followed by the publication of final opportunity maps. This resource was utilized by the provincial government when considering applications for aquaculture, allowing the potential user conflicts at the proposed sites to be identified and minimized.

In Norway there is increasing pressure for activities in the coastal zone to give larger profits, in correspondence with strong international economic competition, as well as national competition between different activities. This requires that the management of the coastal zone allocates rights to resources to the most valuable sector and therefore influences how aquaculture and tourism compete for space. Aquaculture is a significant economic sector for Norway, but its spatial development is partially controlled by the state through the issuing of production licenses by region, encouraging distribution that considers the needs of each region and the capacity of the receiving environment.

Some states have been more specific in their marine planning, explicitly recognizing the interactions between tourism and aquaculture. The Ministry of Reconstruction in Turkey recently developed a coastal zone master plan. It is thought that conflicts with aquaculture could be resolved through the production of marine spatial planning for aquaculture development in suitable areas, with guidelines regarding separation distances. Following are criteria and minimum distances between various types of aquaculture installation and tourism development (dependent on topography, concealment and screening):

- Shellfish aquaculture should be at least 1 km from tourist hotels and holiday homes.
- Cage aquaculture, hatchery and tank farms should be at least 1 km from tourist centres.
- In scenic areas, cage aquaculture, hatchery and tank farms should be kept at distances of 0.5 km, 0.75 km and 1 km respectively.
- Hatcheries, ponds and tank farms should be screened from view with trees and shrubs.
- Cage and raft culture should be restricted in heavily used recreational waters as a safety precaution but should be permitted in waters with light, irregular traffic.
- Fish farm operators should be encouraged to clearly mark the boundaries of farm sites.
- Hatcheries should have their intakes and outfalls clearly buoyed.
- Tourists should not be permitted within farm and hatchery areas.
- Tourism development should be subject to the same controls relating to pollution of the sea as other industries.

CASE STUDY: SOUTH AFRICA, FLOATING FISH FARM IN ALGOA BAY

Introduction

The proposed development of a floating fish farm (marine aquaculture) in Algoa Bay, South Africa, just 2.4 km off the City of Port Elizabeth's lucrative beachfront, was put on hold pending further feasibility studies as directed by the Department of Environmental Affairs for consideration after 134 notices of intention to appeal were submitted to the Minister. An online petition against the proposed development at its approved location generated approximately 17,000 signatures.

Stakeholders stated that whilst they understood the value and need for fish farms in the future with regard to job creation, the concern remained that the current location would have an immediate and potentially disastrous negative impact on the tourism sector. The proposed fish farm, stretching across popular beaches including Hobie and Pollock, could potentially pollute the area and spoil the visual beauty of the seafront. Further, stakeholders stated that should the fish farm find suitable alternative locations, which they were advised were available, then the tourism industry and fish farm could coexist in that destination, creating greater opportunities and income.

The impact of a floating fish farm on the Ironman SA event

For over three decades the Ironman World Championship has brought the world's best athletes together in competition. The inaugural 'Hawaiian Ironman Triathlon' was conceptualized in 1977 as a way to challenge athletes who had seen success at endurance swim, running and biathlon events. Honolulu-based Navy couple Judy and John Collins proposed combining the three toughest endurance races in Hawaii into one event (the 2.4 mile Waikiki Roughwater Swim, the 112 mile Around-O'ahu Bike Race and the 26.2 mile Honolulu Marathon). The Ironman World Championship centers on the dedication and courage exhibited by participants who demonstrate the Ironman mantra that, 'Anything is Possible'.

In a public statement, the Ironman SA management team warned that should the floating fish farm development go ahead then they would move this high-profile event to another destination. The economic impact and loss of revenue to the two host cities in the Eastern Cape Province of South Africa would be significant. Their concern, and supported by all coastal and ocean stakeholders in the province, was that the fish farm could attract great white sharks to an area where the triathletes swim. Generally, great white sharks are in transit during their transoceanic journeys. However, being curious creatures and in some cases familiar with chumming in the vicinity of cages used for cage diving, it is quite possible that they could be drawn to the floating fish farm cages. The risk to triathletes and recreational bathers using popular beaches was considered to be too high.

A new 3-year deal signed between Standard Bank and Ironman SA will have a massive positive financial impact on the metropolitan cities of Port Elizabeth and East London, bringing in millions of extra rand in much-needed revenue to both cities. The annual direct spend in Port Elizabeth every year is in the region of R80 million and around R40 million for East London. The management of the Ironman event predict that direct annual spend in the greater Nelson Mandela Bay metropolitan city (incorporating the City of Port Elizabeth) will increase to around R100 million. As a result of the new partnership with Standard Bank, the World Triathlon Corporation, owner of the Ironman brand, has elevated the status of the Ironman SA in Port Elizabeth to a continental championship race, with the event becoming known as the Standard Bank Ironman African Championship, hosted by Nelson Mandela Bay. This places the event as one of the top six Ironman events in the world. With this change comes a prize purse of $150,000 for professionals, increased world championship qualifying points and 75 amateur qualifying slots to the Ironman World Championship in Kona, Hawaii.

Participation in Ironman SA has grown from just 786 entrants in 2005 to 2371 in 2014, which is a 202% increase. The biggest age group in the 2014 Ironman event is 40-44 years and almost half of all participants are first timers. For the Half Ironman in East London, participation has climbed from 864 entrants in the inaugural 2008 event to 2926 entrants in the 2015 event, an increase of 239% since the event started. Over the years, the Ironman management team has raised R8.7 million for Eastern Cape charities. With Standard Bank on board, they believe it will assist them even more in their quest to continue supporting disadvantaged children with the message that *anything is possible.*

At the time of writing the Department of Fishing & Forestry (DAFF) were exploring alternative sites, such as the area known as Algoa Zone 5, as the preferred location for the floating fish farm project.

Conclusion

The Algoa Bay floating fish farm case study illustrates that when feasibility studies are conducted, with regard to development projects in the ocean, then a cost–benefit analysis and risk assessment should be undertaken to evaluate the economic impact of one development on another. In the case of Algoa Bay, the total value of the fishing industry in the Bay generates about R38 million a year compared to 5 km of beachfront which generates about R3 billion rand a year in tourist revenue (from hotels, holiday flats, restaurants, shops, cultural and sporting events, festivals, etc). Furthermore, the sustainability of a new development should be taken into consideration, for example how many floating cages are required to operate a viable fish farming business. To put at risk a viable and sustainable coastal tourism industry on the assumption that 38 visible cages would provide a good return on investment would not make economic sense. The terms of reference inviting bid proposals should clearly state that bidders must prepare business plans indicating the viability of their project proposal.

Discussion

Undoubtedly the best way to avoid conflicts of interest in the ocean space, and to avoid one ocean development from either displacing or putting at risk another, is to map the spatial area according to zones identified for appropriate investment projects. This requires a marine spatial plan (MSP). This is another example of the need for collaboration because a MSP should be a consultative process whereby all local coastal and ocean stakeholders are provided with an opportunity to participate in the planning process according to their specific perspectives or fields of specialization. In this regard, oceanographers understand wave motion and currents under extreme conditions; scuba divers can locate diving hotspots and marine biodiversity sensitive areas which require protection; and yachting and boating enthusiasts can identify the best places for water-based marine activities and regattas. All of these stakeholders can greatly assist in mapping the underwater world within a spatial area (e.g. zoning). There is not likely to be a single structure that would be fully representative of all these different interests, but they could be easily coordinated by a maritime cluster.

References and further reading

Alongi, D. (2002). 'Present state and future of the world's mangrove forests'. *Environmental Conservation*, 29(3): 331–349.

Alongi, D. (2009). 'Paradigm shifts in mangrove biology' in: G.M.E. Perillo, E. Wolanski, D.R. Cahoon and M.M. Brinson (eds), *Coastal Wetlands: An integrated ecosystem approach*. Cambridge, MA: Elsevier.

Bouillon, S., Borges, A.V., Castaneda-Moya, E., Diele, K., et al. (2008). 'Mangrove production and carbon sinks: A revision of global budget estimates'. *Global Biogeochemical Cycles*, 22.

Bouillon, S., Koedam, N., Raman, A.V. and Dehairs, F. (2002). 'Primary producers sustaining macro-invertebrate communities in intertidal mangrove forests'. *Oecologia*, 130: 441–448.

Bouillon, S., Moens, T., Koedam, N., Dahdouh-Guebas, F., et al. (2004). 'Variability in the origin of carbon substrates for bacterial communities in mangrove sediments'. *FEMS Microbiology Ecology*, 49: 171–179.

Bouillon, S., Mohan, P.C., Sreenivas, N. and Dehairs, F. (2000). 'Sources of suspended organic matter and selective feeding by zooplankton in an estuarine mangrove ecosystem as traced by stable isotopes'. *Marine Ecology Progress Series*, 208: 79–92.

Dempster, T. and Sanchez-Jerez, P. (2008). 'Aquaculture and coastal space management in Europe: An ecological perspective' in: M. Holmer, K. Black, C.M. Duarte, N. Marbà, et al. (eds), *Aquaculture in the Ecosystem*. New York: Springer International Publishing AG.

FAO (Food and Agriculture Organization) (2012). 'The state of the world fisheries and aquaculture 2012'. www.fao.org/3/a-i2727e.pdf (accessed: May 2017).

FAO (Food and Agriculture Organization) (2014). 'The state of world fisheries and aquaculture: Opportunities and challenges'. www.fao.org/3/a-i3720e.pdf (accessed: April 2017).

Grant, A. (2011). 'The siting and design of aquaculture in the landscape: Visual and landscape considerations'. Prepared for Scottish Natural Heritage. www.snh.org.uk/pdfs/publications/heritagemanagement/marineaquaculture.pdf (accessed: April 2017).

Hauraki Gulf Marine Park (August 2014). 'Aquaculture round-table Report 1: Collation of background information' in *Hauraki Gulf Marine Spatial Plan*. www.seachange.org.nz/PageFiles/387/Aquaculture%20Roundtable%20Report%201%20Collection%20of%20background%20information.pdf (accessed: April 2017).

Kristensen, E. (2008). 'Mangrove crabs as ecosystem engineers; with emphasis on sediment processes'. *Journal of Sea Research*, 59: 30–43.

Kristensen, E., Bouillon, S., Dittmar, T. and Marchand, C. (2008). 'Organic carbon dynamics in mangrove ecosystems: A review'. *Aquatic Botany*, 89: 201–219.

Lee, S.Y., Primavera, J.H., Dahdouh-Guebas, F., McKee, K., et al. (2014). 'Ecological role and services of tropical mangrove ecosystems: A reassessment'. *Global Ecology and Biogeography*, 23: 726–743.

Miranda, S.J.M., Reyes, J.O., Tan, A., Villanueva, L.C., et al. (2013). 'Mangrove plantation as a tourist attraction in San Juan Batangas, Philippines'. *International Journal of Academic Research in Business and Social Sciences*, 3(8): 1–5.

de Monbrison, D. (2004). 'Mediterranean marine aquaculture and environment: Identification of issues'. In provision of the IUCN workshop on Mediterranean Aquaculture and Environment. www.uicnmed.org/web2007/documentos/Mediterranean_marine_aquaculture.pdf (accessed: May 2017).

Morales, M.I., Masagca, J.T., Araojo, A.E. and Vargas, S.R. (2014). 'Coastal and mangrove eco-tourism in Catanduanes Island (Philippines): A menace or a bonus?' International Conference on Latest Trends in Food, Biological & Ecological Sciences: Phuket, Thailand. http://iaast.org/upload/6185A0714028.pdf (accessed: April 2017).

Nimmo, F., Cappell, R. and Grant, A. (2009). 'Literature review of evidence that fish farming impacts on tourism'. Prepared by Royal Haskoning for Scottish Aquaculture Research Forum. www.sarf.org.uk/cms-assets/documents/28824-626142.sarf045-tourism-appendix-a-final.pdf (accessed: April 2017).

Pérez, O.M., Telfer, T.C. and Ross, L.G. (2005). 'Geographical information systems-based models for offshore floating marine fish cage aquaculture site selection in Tenerife, Canary Islands'. *Aquaculture Research*, 36(10): 946–961.

Pérez, O.M., Telfer, T.C., Beveridge, M.C.M. and Ross, L.G. (2002). 'Geographical information system (GIS) as a simple tool to aid modelling of particulate waste distribution at marine fish cages sites'. *Estuarine, Coastal and Shelf Science*, 54: 761–768.

Primavera, J.H. (2006). 'Overcoming the impacts of aquaculture on the coastal zone'. *Ocean & Coastal Management*, 49(9–10): 531–545.

Sousa, W.P. and Dangremond, E.M. (2011). 'Trophic interactions in coastal and estuarine mangrove forest' in: E. Wolanski and D.S. McLusky (eds), *Treatise on Estuarine and Coastal Science*. Waltham, MA: Elsevier, p.43–93.

Stead, S. (2015). 'Mariculture: Aquaculture in the marine environment' in: H.D. Smith, J.L.S. de Vivero and T. Agardy (eds), *Routledge Handbook of Ocean Resources and Management*. London and New York: Routledge.

Stephanou, D. (1998). 'Marine aquaculture development and tourism: The case of Cyprus.' *Cahiers Options Méditerranéenes*, 43: 35–40.

Technical Centre for Agriculture & Rural Cooperation (CTA) (2013). 'Fish-farming: The new driver of the blue economy?' www.cta.int/en/article/2013-06-24/fish-farming-the-new-driver-of-the-blue-economy.html (accessed: April 2017).

Wijkström, U.N. (2012). 'Is feeding fish with fish a viable practice?' in: R.P. Subasinghe, J.R. Arthur, D.M. Bartley, S.S. De Silva, et al. (eds), *Proceedings of the Global Conference on Aquaculture 2010*, Phuket, Thailand, p.33–55.

Chapter 6

Marine renewable energy in the context of sustainable oceans

Offshore drilling for oil or gas is viewed as a development that carries with it massive risk to the environment and the economic drivers that rely on healthy and accessible oceans and coasts. Coastal tourism and thriving fisheries are at risk from offshore drilling accidents. As Gulf Coast states painfully learnt in the 2010 Deepwater Horizon disaster, offshore drilling accidents can devastate robust tourism industries, the health of marine ecosystems, and both the productivity and the reputation of the marine fisheries supported by those ecosystems. There are those who are convinced that the oceans could shape the energy future by pursuing offshore wind and wave energy and development, rather than drilling for oil.

Offshore wind farms and wave and tidal projects could have complimentary impacts on birds, fish, marine mammals, and marine and coastal ecosystems, while also supporting sustainable ocean industries such as fisheries and tourism. Ongoing monitoring and modification of offshore wind farms, current, tidal and wave projects, and related grid infrastructure can positively address biodiversity and ecosystem concerns through the construction, authorization, operation and maintenance, and the abandonment and reclamation of these facilities. Prioritizing development and protection of existing coastal-dependent industries like tourism and fishing, while boosting support for renewable energy development, is a common-sense approach; but it requires a high level of stakeholder collaboration when evaluating the pros and cons.

Renewable energy and sustainable ocean development

A recent PricewaterhouseCoopers analysis of the tourism industry in Virginia, US, reported that the sector supports more than 200,000 jobs, yielding an economic impact of more than US$20 billion in 2011, and that Virginia's beaches alone attracted nearly 10% of the state's tourists. Virginia's coast and ocean also support thriving fisheries; in 2011 fishermen landed 247,000 tons of seafood in Virginia, worth more than US$191 million, ranking it the third largest seafood producer in the US by weight. Another vital activity carried out in Virginia is

Naval Station Norfolk, the largest naval complex in the world. One of the state's primary economic engines, the naval station supports more than 71,000 military and civilian employees, and, in 2009, was responsible for nearly US$15 billion in economic impact in Virginia.

In 2010 the US Department of Defense determined that 74% of the potential offshore areas for oil and gas leasing would directly interfere with the extensive military activities that are carried out in the region, including ordnance training and aircraft carrier operations. A Virginia representative stated that the Navy's operations were much more important to the Virginia economy, let alone national security, than drilling operations. Furthermore, most of the areas where the Navy did not find potential conflicts were either in major shipping lanes or off the northern part of the coast where Virginia's desire to drill would come into conflict with their neighbours in Maryland where the sentiment is distinctly anti-drilling. (After watching oil gush into the Gulf of Mexico unchecked for more than a month, Maryland Governor took an unequivocal stance, saying that he would be opposed to any drilling off the Chesapeake Bay.)

Virginia can still look to the ocean to shape its energy future, but rather than drilling for oil, the state can pursue offshore wind energy development. A campaign website noted that Virginia's offshore wind resource could generate up to 3,200 megawatts of electricity, and the development could support as many as 11,600 jobs. Furthermore, the Department of the Interior's 'Smart from the Start' program has already identified sites for wind turbines off Virginia's coast that would not conflict with either military operations or the shipping industry. Offshore wind holds vast promise for producing clean energy at modest prices that will be stable over the long term (much more than can be said for oil or natural gas).

Rather than looking to the industries of the past and falling victim to another black gold rush that would compromise Virginia's robust blue economy, Virginia's leaders agree that they should instead look to a proven source of home-grown energy development that will create new jobs without jeopardizing current ones or contributing to the nation's unsustainable thirst for climate-altering fossil fuels. Prioritizing development and protection of the state's existing coastal-dependent industries like tourism and fishing, and its world class military emplacements, while boosting support for renewable energy development is a common-sense approach to moving one of America's most historic states toward a brighter economic future.

Ocean energy

Forever moving, our restless oceans have enough energy to power the world (for a review, see O'Neill, Elefant and Agardy 2015). As long as the Earth turns and the moon keeps its appointed cycle, the oceans will absorb and dissipate vast amounts of kinetic energy – a renewable energy resource of enormous potential. However, harnessing this resource has proven more difficult than first thought. One of the instalments of the GLOBE-Net Series on Renewable Energy suggested that we should be looking at how the power of the oceans might eventually find its place among other forms of renewable energy. According to the UN, 44% of the world's population lives within 150 km of an ocean coast. In Canada and Australia the number is much higher at 80%. In the US, 53% of the population live in close proximity to an ocean. Thus it is only natural that many countries look to the oceans as a source of energy to be harnessed. How they seek to exploit this resource varies according to factors of geography and available technologies. The two main forms of energy associated with our oceans are tidal power and wave power, born of the same source but different in how they turn energy into electricity.

Tidal power

Tidal power coverts the energy of tides into electricity utilizing the rise and fall of the ocean tides. The stronger the tide, either in water level height or tidal current velocities, the greater the potential for tidal electricity generation. Tidal generators act in much the same way as do wind turbines. However the higher density of water (832 times that of air) means that a single generator can provide significant power at velocities much lower than those associated with the wind power generators.

Tidal stream systems and barrages are the two main categories of tidal power. Barrages are similar to hydroelectric dams but are placed in an estuary bay or river mouth, where they act as barriers that create artificial tidal lagoons. When water levels outside the lagoon change as relative to water levels inside, turbines in the barrages are able to produce electrical power. There are only three such structures in the world: the Rance River in France, Canada's Bay of Fundy and Kislaya Guba in Russia. Tidal stream systems simply rely on individual turbines which are placed in the water column; moored to be suspended, floating or anchored to the ocean floor. As the tide flows in or out, electrical energy is produced as water moves through the turbine.

Tidal power boasts several advantages over other types of renewable energy technology because tides are more predictable and reliable than wind energy or sunny days for solar power. Tidal energy has an efficiency ratio of approximately 80% in terms of converting the potential energy of the water into electricity. Tidal stream system turbines are only a third of the diameter of wind rotors of the same power output. As with wind power, location is an important factor in terms of being able to harness the earth's natural energy. Tidal stream systems must be located in areas with fast currents where natural flows are concentrated between natural obstructions, for example at the entrances to bays and rivers, around rocky points or headlands, or between islands and other land masses.

Wave power

Ocean surface waves are a considerable source of energy potential, but energy that is not as restricted in terms of location as tidal energy systems. Typically wave energy is captured using buoys which generate mechanical energy as they oscillate vertically from wave motion. Terminator devices extend perpendicular to the direction of wave travel and capture or reflect the power of the wave. Water enters through a subsurface opening into a chamber with air trapped above it and wave action causes the captured water column to move up and down like a piston to force the air though an opening connected to a turbine. A point absorber is a floating structure with components that move relative to each other due to wave action (e.g. a floating buoy inside a fixed cylinder). The relative motion is used to drive electro-mechanical or hydraulic energy converters. Attenuators are long multi-segment floating structures oriented parallel to the direction of the waves. The differing heights of waves along the length of the device causes flexing where the segments connect, and this flexing is connected to hydraulic pumps or other converters.

Overtopping devices have reservoirs that are filled by incoming waves to levels above the average surrounding ocean. The water is then released, and gravity causes it to fall back toward the ocean surface. The energy of the falling water is used to turn hydro-turbines. Wave power varies considerably in different parts of the world, meaning wave energy cannot be harnessed effectively everywhere. According to the Ocean Renewable Energy Group, a Vancouver-based organization that promotes the development of ocean energy in Canada, regions considered to have 'good' wave energy resources are generally those found within 40 to 60 degrees of latitude, where the strongest winds are found. Wave-power rich areas of the world include the western coasts of Scotland, northern Canada, southern Africa, Australia and the northwestern coast of the US.

The challenges

Both ocean wave and tidal power have attracted growing interest from investors and power utilities looking for the next long-term play in renewable energy. Using environmentally responsible technologies, we have a tremendous opportunity to harness energy produced from ocean waves, tides or ocean currents, free-flowing water in rivers and other water resources, to provide clean and reliable power.

Despite this enormous potential of ocean energy, there remain many challenges that have proven difficult to overcome, and which explain why ocean energy remains the least developed of all forms of renewable energy. Problems still exist regarding cost, maintenance, environmental concerns and our still imperfect understanding of how power from the oceans will impact on the world's energy infrastructure. For example, turbines are susceptible to bio-fouling (the growth of aquatic life on or in the turbine). This can severely inhibit the efficiency of energy production and is both costly and difficult to remove. Turbines are also prone to damage from ocean debris. Finally, turbines may be hazardous to marine life: the impacts on marine life are still largely unknown, but concern is warranted.

Barrage systems are affected by problems of high initial infrastructure costs associated with construction and reducing the resulting environmental problems.

For example, independent research on the economics of building the proposed Severn Barrage in the UK revealed that, taking environmental costs into account, the structure could cost as much as £12 billion to create. Barrage impacts include a decrease in the average salinity and turbidity within a barrage, which significantly alters associated ecosystems. Wave power systems present their own set of challenges. Most electric generators operate at higher speeds, and most turbines require a constant, steady flow. Unfortunately wave energy is slow and ocean waves oscillate at varying frequencies. The rough realities of the marine environment have also proven difficult to deal with, especially for companies seeking to remain cost effective. Constructing wave devices that can survive storm damage and saltwater corrosion add to development costs.

The future

Modern advances in ocean energy technology may eventually see large amounts of power generated from the ocean, especially from tidal currents using tidal stream designs. The technology is still in its infant stage and most projects that exist or are in development stages are mainly pilot projects. Yet the promise remains, despite being less established than energies such as solar, thermal, wind and biomas. According to the CEO of Pelamis Wave Power: as the technology develops and becomes more affordable, which it will over time, we can continue to expand pretty much anywhere where there is an ocean.

The market potential for tidal power still remains unclear. Sector analysts believe initial public offerings for wave and tidal power projects will be much harder to price than for comparable wind power projects, because wave firms cannot give exact estimates on the scale of benefits and few have technologies that are up and running. According to research studies, ocean power cannot match fossil fuels for electricity production, but could be competitive with other forms of renewable energy.

CASE STUDY: ENGLAND, POOLE HARBOUR

Overview

Global warming and its impact on the environment, society, economies and security is one of the fundamental concerns of our time. In response, the UK government has put in place a legally binding target of an 80% reduction in greenhouse gas emissions on 1990 levels by 2050. On the south coast of England, Poole Harbour has been recognized as a potential, highly predictable source of tidal and heat energy. Local groups embarked on a feasibility study for harnessing this energy for the benefit of the community.

The purpose of the feasibility study was to examine the potential conflict of interest between the laudable aims of promoting the use of renewable energy and of safeguarding ecosystems and their biodiversity. The Poole Harbour feasibility study could therefore be used as a case study to consider the environmental and economic costs and benefits of a community renewable energy (CRE) project (the Poole Tidal Energy Partnership). In the context of an area subject to a number of statutory and non-statutory designations to protect nationally and internationally important habitats and species, the case study could identify key environmental legislation, including spatial planning law and policy, which will facilitate exploring whether there is potential for reconciling what may be perceived as competing objectives for sustainable development.

Poole Harbour is the one of largest natural saltwater harbours in the UK. It represents a highly predictable tidal and heat energy potential that has prompted local interest in investigating the possibility of harnessing it for the benefit of the area through community schemes. At the same time, Poole Harbour has a variety of marine, freshwater and terrestrial habitats which, together with surrounding Areas of Outstanding Natural Beauty, are subject to a number of statutory and non-statutory designations. The proposal to harness marine renewable energy from Poole Harbour will provide an insight into the different challenges faced by CRE schemes. These include ecological, legal, economic and planning issues that need to be considered in order to achieve climate change mitigation without loss of biodiversity.

The Poole Tidal Energy Partnership

The Poole Tidal Energy Partnership (PTEP) was formed on 4 November 2011 in response to local policies on renewable energy. This partnership, with participants from the Borough of Poole, Transition Town Poole and Bournemouth University, combines local knowledge, community representation and technical expertise to form a single organization focused upon driving forwards the vision of a local, secure and green energy supply. PTEP's aim is to provide renewable energy at affordable prices to local residents whilst encouraging a high degree of community participation in the project.

The economics of PTEP

As a fundamental concern to both society and the principle of sustainable development, it was important for the scheme at Poole Harbour to consider the economic benefits and disadvantages of development. Significant opportunities exist for the UK to make the most of its island geography and harness the power of its coastline in the production of low carbon energy.

In November 2011, the Government launched the Renewable Heat Incentive (RHI) scheme to provide 'financial support to non-domestic renewable heat generators' (House of Commons Library 2017). This seeks to provide a 20 year income stream to those who install a system covered under the scheme with the aim of making renewable heat an attractive commercial option. Renewable heat is simply any heat generated from a renewable source and will include the use of ground- and water- (marine-) source heat pumps.

When evaluating such a project purely from an investor perspective, giving consideration only to direct, monetary values, Woodruff (2007, p.18) states that 'a project is financially viable when a project's revenues exceeds its costs'. Clearly, economic analysis from a societal perspective provides a much wider scope of factors to be considered and may lead to contrasting results. Using this economic approach, Woodruff (2007) advised 'the project with the greatest net benefits should be chosen among all technologically feasible options for providing electricity'. Considering an economic approach to the evaluation of renewable energy projects, key non-monetized costs and benefits may need to be incorporated into the analysis.

However, the difficulty here is that these costs and benefits are difficult to quantify. Air or noise pollution during construction may be a consideration but perhaps the greatest non-monetized cost is the potential effects on the ecosystem within the harbour and beyond. The impact on and valuation of ecosystem services is a consideration which must form part of the appraisal of future energy projects in general, and especially in a project where the site is as significant as Poole Harbour. These non-monetized costs may be countered by non-monetized benefits, such as macroeconomic benefits for the UK as a whole through the use of renewable energy and contribution towards EU energy targets (DECC 2015). However, whether this leads to a net benefit is a subject of much debate.

Poole Harbour ecology

As noted, Poole Harbour is an area of significant environmental importance. In order to appreciate the impact that the PTEP project will have on Poole Harbour, an understanding of the geological features and ecological importance of the area is imperative. Poole Bay is a drowned valley that was formed at the end of the last ice age. The bay is an estuary formed from a number of rivers, the largest of which is the River Frome, and is surrounded by shallow creeks and islands. Poole Harbour, located to the west of Poole Bay, is one of the largest natural harbours in the UK, and has a shallow water depth of approximately 1 to 5 m. Twice a day, between 30,400 to

45,000 cubic metres of water enters and leaves Poole Harbour, offering huge potential for the implementation of marine renewable projects.

Another relevant feature of the Harbour is the unique intertidal habitats which have resulted in the area being internationally important for populations of breeding and migratory bird species. The Harbour is a heavily utilized and managed site covering an area of approximately 3600 ha with the densely populated north side of the harbour serving the town of Poole. There are local management schemes that aim to keep disturbance in the area to a minimum, particularly in bird sensitive areas between Studland Bay, at the entrance of the Harbour, and Keysworth Point, to the west. The Poole Harbour Aquatic Management Plan, overseen by the Poole Harbour Steering Group, advocates zoning, 'to assist in the reduction of disturbance to environmentally sensitive areas'. The plan's main objective is 'to promote the safe and sustainable use of Poole Harbour' whilst balancing the competing demands of user groups (Poole Harbour Aquatic Management Plan 2011).

Due to its recognized ecological importance, any development in Poole Harbour needs to take into consideration certain international conventions, European and UK marine and nature conservation laws, as well as water quality legislation. This results in a complex environmental legal framework that gives several layers of protection, sometimes from different points of views, to the same site.

Wetlands

Wetlands are defined as 'areas of marsh, with water that is static or flowing, fresh, brackish or salt, including areas of marine water whose depth does not exceed 6 m'. Due to their ecological importance, parts of Poole Harbour were designated a RAMSAR wetland site in 1999. In England all terrestrial areas included within listed RAMSAR sites are currently Sites of Special Scientific Interest (SSSI). They are legally protected under the Wildlife and Countryside Act 1981 as amended. Poole Harbour was first designated as a SSSI in 1964 under the National Parks and Access to the Countryside Act 1949 to protect its intertidal and coastal habitats. Poole Harbour was designated because it supports species that are rare or unique.

According to The Conservation of Habitats and Species Regulations 2010, compensation must be of equal or higher quality than the habitat it is replacing. Obviously, this is no easy feat and many studies have been undertaken to understand the implications of habitat compensation. This is relevant to Poole Harbour's distinct environment which provides important habitats for vulnerable birdlife.

Spatial planning

It is important to recognize that a key aspect to the implementation of any development project is the understanding of the legal framework, plans and policies for the control or management of development. The merits or otherwise of a proposal will be examined and assessed with regard to policies and plans, and determined by the granting or refusal of consent. In the context of coastal or marine works, the two main types of development consent required are:

- Planning permission for infrastructure to the boundary jurisdiction of the local planning authority (generally above the Mean Low Water Mark (MLWM)) under the Town and Country Planning Act 1990.
- A Marine Licence for works below Mean High Water Springs (MHWS) under the Marine and Coastal Access Act 2009.

The two systems work independently of each other and are administered by local planning authorities and the Marine Management Organization (MMO) respectively. There is, however, a reciprocal system of consultation on development proposals. Applications for development are determined in accordance with the Local Plan of the local planning authority in the case of land-based development and in accordance with the relevant Marine Plan for offshore development. In locations where Marine Plans have yet to be produced, in the interim, direction on licensing decisions will be given by the Marine Policy Statement (MPS). Focusing on the inshore development of the Poole Harbour project, the construction, alteration or improvement of any works either in or over the sea or on or under the sea bed (including offshore generating stations with a capacity between 1 and 100 megawatts) requires a marine licence.

Stakeholder engagement

Environmental management (and decision making) is as much about managing human activities as managing land and water, therefore requiring the effective engagement of all stakeholders (including government agencies, NGOs, the business community and the general public) in the process of information gathering and policymaking. Engagement and participation engenders a sense of ownership of information and policymaking resulting in a common understanding of issues and a negotiated learning process. In the UK, as in many countries, part of the political agenda for improving governance at national and local level is the promotion of community capacity building across a range of policy areas including spatial planning.

Participation is an integral part of strategic environmental assessment (SEA) and environmental impact assessment (EIA Directive) and in many countries, including the UK, participation is a mandatory requirement for all planning decisions. Both the established terrestrial planning system and the emerging marine planning system seek to engage the local community and stakeholders at each stage of the planning process: from information gathering, plans and policymaking, to implementation. Both systems incorporate a Statement of Community Involvement or Statement of Public Participation which set out how local communities and stakeholders will be consulted, how they can be involved in the formulation, alteration and review of policies, and the part they can play in consideration of applications for development.

Conclusion

In principle, community renewable energy projects, such as the PTEP initiative, are an innovative and successful way to supply energy to homes and businesses on a small scale with a range of associated benefits for local people and ecosystems. These types of projects require detailed consideration of financial and technological viability, as well as their potential impacts on the environment. Environmental impacts assume even greater importance when the environment into which the development is introduced is valued for its habitats, species and amenity, particularly if it is subject to statutory designations and legal protection, as is the case with significant areas of Poole Harbour.

Given the government's twin objectives of mitigating climate change through the promotion of renewable energy and sustaining and enhancing biodiversity, it is crucial to consider the capability of decision makers to integrate or reconcile these two policy objectives. The question is not whether one policy objective should take precedence over another, but whether spatial planners can work with engineers, ecologists, economists, environmental lawyers, the public and other stakeholders to gather robust evidence to realize renewable energy opportunities with minimal impact to species, habitats or ecosystems.

The Poole Harbour case study is an excellent example of what can be achieved by local authorities and communities working together via a consultative process, especially when there are competing objectives for sustainable development. The Poole Harbour renewable energy project in the southwest of England involved three councils (Dorset County Council, Borough of Poole Council and Bournemouth Borough Council). These councils, together with local stakeholders and universities, developed a strategy known as the Renewable Energy Strategy to 2020 (Dorset Energy Partnership: DEP 2013), which aims to harness viable renewable energy resources to maximize local economic, environmental and community benefits.

Discussion

Just as land-use planning seeks to reconcile competing claims for the use of land, the marine planning system seeks to 'contribute to the achievement and integration of sectoral/activity specific policy objectives within a framework of economic, social and environmental considerations' (MPT 2011). With reference to Marine Plans, which provide a locally relevant expression of policy, implementation and delivery, a key principle will be to promote compatibility and to reduce conflict. During the consultation process in response to the comments received by respondents on implicit policy priorities, the government needs to make clear that the MPS does not prioritize one activity over another. Furthermore, relative priorities are most appropriately determined through the marine planning process, taking into account a wide range of factors alongside UK policy objectives, including the specific characteristics of individual areas.

On the question of clarity on the management of conflict, Marine Plans should identify areas of constraint and areas of opportunity for a range of activities. By engaging local communities and stakeholders in all stages of the planning process, support for plan-led policies is engendered, opportunities for compatible uses are identified, and potential conflicts are minimized. Further, the role of evidence in conflict resolution is emphasized. Marine spatial planning is a public process of analyzing and allocating the spatial and temporal distribution of human activities in marine areas to achieve ecological, economic and social objectives that usually have been specified through a political process. Characteristics of marine spatial planning include ecosystem-based, area-based, integrated, adaptive, strategic and participatory.

Marine spatial planning is not an end in itself, but a practical way to create and establish a more rational use of marine space and the interactions among its uses, to balance demands for development with the need to protect the environment, and to deliver social and economic outcomes in an open and planned way.

References and further reading

Coyle, M. and Wiggins, S. (2010). 'European marine site risk review'. Prepared for Natural England. http://publications.naturalengland.org.uk/file/50076 (accessed: May 2017).

DECC (UK Department of Energy & Climate Change) (2015). 'Delivering UK energy investment: Low carbon energy'. http://citeseerx.ist.psu.edu/viewdoc/download?doi=10.1.1.699.4952&rep=rep1&type=pdf (accessed: May 2017).

DEP (Dorset Energy Partnership) (2013). 'Bournemouth, Dorset and Poole renewable energy strategy to 2020'. www.bournemouth.gov.uk/environment-and-sustainability/ClimateChangeandCarbonReduction/Documents/BournemouthDorsetandPooleRenewableEnergyStrategyto2020.pdf (accessed: April 2017).

Dorset Area of Outstanding Natural Beauty (2012). 'Annual review 2010–2011'. www.dorsetaonb.org.uk/assets/downloads/Annual_Review/2010-2011%20annual%20review%20a4%20print%20version.pdf (accessed: April 2017).

Edenhofer, O., Pichs-Madruga, R., Sokona, Y., Seyboth, K., et al. (2011). *Renewable Energy Sources and Climate Change Mitigation*. Special Report of the Intergovernmental Panel on Climate Change. Cambridge: Cambridge University Press.

Ginige, T. (2013). 'Harnessing marine renewable energy from Poole Harbour: A case study'. Work in Progress. Bournemouth, UK: School of Applied Sciences, Bournemouth University.

House of Commons Library (2017). 'Renewable heat incentive'. Briefing paper number 6328. http://researchbriefings.files.parliament.uk/documents/SN06328/SN06328.pdf (accessed: May 2017).

Julius, S. (2012). 'Marine renewable energy feasibility: Poole Harbour case study'. Dissertation. Bournemouth, UK: School of Applied Sciences, Bournemouth University.

MPT (Marine Planning Team) (2011). 'A description of the marine planning system for England'. http://webarchive.nationalarchives.gov.uk/20121204124616/http:/archive.defra.gov.uk/corporate/consult/marine-planning/110318-marine-planning-descript.pdf (accessed: May 2017).

O'Neill, S., Elefant, C. and Agardy, T. (2015). 'Renewables: An ocean of energy' in: H.D. Smith, J.L.S. de Vivero and T. Agardy (eds), *Routledge Handbook of Ocean Resources and Management*. London and New York: Routledge.

Poole Harbour Aquatic Management Plan (amended 2011). 'Non-technical summary'. www.pooleharbouraqmp.co.uk/ (accessed: April 2017).

Woodruff, A. (2007). 'An economic assessment of renewable energy options for rural electrification in Pacific Island countries'. Prepared for Pacific Community Geoscience Division. https://pdfs.semanticscholar.org/24e8/2b1ad53567b6be18d3f56887424158825296.pdf (accessed: April 2017).

Chapter 7

Integrated coastal zone management

Integrated coastal zone management (ICZM) is a strategy for an integrated approach to planning and management, in which all policies, sectors and, to the highest possible extent, individual interests are properly taken into account, with the proper consideration given to the full range of temporal and spatial scales, and involving stakeholders in a participative way. It demands good communication among governing authorities (local, regional and national), and promises to address all three dimensions of sustainability: socio/cultural, economic and environmental. This comprehensive strategy thus provides management instruments that are not *per se* included or foreseen in the different policies and directives. Maritime clusters facilitate the implementation of ICZM plans by creating opportunities for collaboration between maritime authorities, and coastal and ocean stakeholders.

UNEP Global Program of Action for the Protection of the Marine Environment from Land-based Activities

In 1995 over 108 governments laid down:

> their commitment to protect and preserve the marine environment from the impacts of land-based activities thereby setting as their common goal sustained and effective action to deal with all land-based impacts upon the marine environment, specifically those resulting from sewage, persistent organic pollutants, radioactive substances, heavy metals, oils (hydrocarbons), nutrients, sediment mobilization, litter, and physical alteration and destruction of habitat.
>
> (UNEP 1995)

The commitment, made under the auspices of the Global Program of Action for the Protection of the Marine Environment from Land-based Activities (GPA) and the associated 1995 Washington Declaration, has been renewed through the Montreal and Beijing declarations in 2001 and 2006 respectively.

The GPA encourages governments and regional organizations to prepare and implement *comprehensive, continuing and adaptive* action plans to protect the marine environment, recognizing the effects on food security, poverty alleviation, and ecosystem health, as well as the resulting economic and social benefits. In parallel, some of the 18 Regional Seas Conventions and Action Plans (RSCAPs) have negotiated legally binding protocols on land-based activities and sources of marine pollution. Governments determined at the Second Intergovernmental Review of the GPA in Beijing in 2006 that the period 2007–2011 would focus on mainstreaming the implementation of the GPA in national development planning and budgetary mechanisms, and indicated that mainstreaming will require integration of the GPA across sectors and ministries and also integration into domestic and international aid budgets, development plans, strategies and actions. The Third Intergovernmental Review on the implementation of the GPA was held in January 2012 in Manila, Philippines, preceded by a Global Conference on Land-Ocean Connections.

UNEP and the GPA

The United Nations Environment Program (UNEP) serves as the Secretariat of the GPA with a mandate to: (1) promote and facilitate its implementation at regional, national and sub-national levels; and (2) play a catalytic role with other organizations and institutions in implementing the GPA internationally. Consistent with the directive of governments and the UNEP Marine and Coastal Strategy, UNEP conducts international, regional and national activities to advance the GPA, under the following categories:

- *Collaboration and coordination:* UNEP collaborates with strategic partners to develop and showcase the 'foundations' for national action on: nutrient management; wastewater management; marine litter; and the physical alteration and destruction of habitat.
- *Policy and normative action:* UNEP makes the case, through demonstration projects, for mainstreaming the protection of coastal and marine ecosystems.
- *Capacity building and technology support:* UNEP provides tools to governments and regional organizations to support decision making and planning.
- *Assessment:* UNEP monitors, and reports on, innovative and best-practice approaches by governments and other stakeholders to protect the marine environment from land-based activities.
- *Awareness raising and outreach:* UNEP increases awareness of coastal degradation and promotes the GPA as a multilateral initiative embodying the principles of ecosystem management where land and sea meet.

The Regional Seas Conventions and Action Plans (RSCAPs) play a fundamental role in implementing the GPA. The governments of some of the RSCAPs have negotiated legally binding conventions/protocols that contribute to achieving its objectives, while others have similarly contributed by negotiating action plans. In some regions, projects funded by the Global Environment Facility (GEF) have resulted in strategic action plans for specific ecosystems, which contribute to the GPA and provide examples of the regional collaboration envisaged by its drafters.

Activities of the Secretariat include:

- UNEP has supported the development and implementation of regional and national action plans across the globe, and an increased use of both integrated coastal area management and environmental impact assessment.
- Legally binding agreements on land-based sources of pollution have been adopted in the Caribbean, Mediterranean and Western Indian Ocean, as have the global conventions on Prior Informed Consent and Persistent Organic Pollutants (POPs). Activities supported by UNEP have catalyzed a number of ratifications to RSCAPs, including their protocols.
- Since 2003, UNEP-GPA has been training municipal wastewater managers on 'Improving Municipal Wastewater Management for Coastal Cities' together with the UNESCO-IHE Institute for Water Education, and other partners around the world. Over 1,350 experts from 67 countries have so far benefited from the training. An evaluation conducted in 2009 indicates that this has had an impact at the local level, which has led to individual projects, research projects, the initiation of country-driven follow-up training courses, and a number of examples of changes in policy guidelines and regulations.
- UNEP has produced a number of guidelines, handbooks, Rapid Response Assessments, and State of the Marine Environment reports, including the 2010 report 'Sick Water? – The central role of wastewater management in sustainable development'. As a follow-up, UNEP has initiated a UN-Water Taskforce on Wastewater Management to develop a comprehensive multi-stakeholder agenda on managing wastewater sustainably.
- Twelve Regional Action Plans for Marine Litter have been developed through partnership with the RSCAPs, and the 2009 report 'Marine Litter: A global challenge' was the first attempt ever to take a worldwide stock of marine litter levels across 12 different regions. Other marine litter publications focus on market-based instruments, survey and monitoring guidelines, and abandoned, lost or otherwise discarded fishing gear.
- UNEP established the Global Partnership on Nutrient Management (GPNM) and acts as its Secretariat. This provides a platform for governments, UN agencies, scientists and the private sector to forge a common agenda, mainstreaming best practices and integrated assessments, so as effectively to render both policymaking and investments 'nutrient proof'.

'Blue Diamonds: Oceans and coasts' is a newsletter to raise awareness amongst stakeholders of GPA issues and activities. The name recognizes that the marine and coastal environment has significant 'value' and is an asset which will return or repay dividends on an investment over time, producing wealth in terms of jobs, trade, livelihoods and ecosystem services like water recycling and shoreline protection. The diamond's strength symbolizes cooperation among the interest groups and stakeholders operating at the land–ocean interface. A diamond also has great beauty, and purity or clarity increases its value.

Integrated coastal management

The management of coastal areas has during the last ten years received considerable attention and got a prominent place on the international agenda in connection with the United Nations Conference on Environment and Development (UNCED) in Rio de Janeiro in 1992 and the Johannesburg World Summit on Sustainable Development in 2002 with the passing of Agenda 21, chapter 17.

Coastal management is the complex interaction of laws, programs and efforts to evaluate tradeoffs and make decisions about how to use, conserve and value the resources and opportunities of the coastal zone. As a society, how do we want our coasts to function, what do we want them to look like, and what uses do we want them to accommodate? ICZM aims to reduce or eliminate such problems, resulting in ethical and economic benefits. Ethical benefits include sustainable development, the promotion of social equity (through consideration of viewpoints of all stakeholders) and protection of traditional uses of coastal resources. Economic benefits accrue from an integrated approach to management which can result in lower costs when compared to management for separate sectors. Effective planning for the future also provides cost benefits.

Sustainable development has been defined as development that meets the needs of the present without compromising the ability of future generations to meet their own needs (WCED 1987). Amongst many definitions, we will define ICZM as:

> A continuous and dynamic process that unites government and community, science and management, sectoral and public interest in preparing and implementing an integrated plan for the protection and development of the coastal ecosystem and resources.
>
> (GESAMP 1996)

ICZM is thus defined as being an iterative and collective process which must be coordinated using a multidisciplinary, multi-tool approach according to a non-linear timescale. Several categories of actors are involved in integrated coastal management. It is a process where we denote many types of integrations:

- Integration between land and sea.
- Integration across sector interests comprising utilization and protection, commercial and recreation interests, etc. (i.e. horizontal integration).
- Integration between authorities at country and municipal levels and public participation (i.e. vertical integration).

The following principles underpin policy in ICZM:

- *National asset:* The coast must be retained as a national asset, with public rights to access and benefits from many opportunities provided by coastal resources, and that resources are used on a sustainable basis for the benefit of present and future generations.
- *Economic development:* Coastal economic development opportunities must be optimized to meet society's needs and to promote the wellbeing of coastal communities.
- *Social equity:* Coastal management efforts must ensure that all people, including future generations, enjoy the rights of human dignity, equality and freedom.
- *Ecological integrity:* The diversity, health and productivity of coastal ecosystems must be maintained and, where appropriate, rehabilitated.
- *Holism:* The coast must be treated as a distinctive and indivisible system, recognizing the inter-relationships between coastal users and ecosystems and between the land, sea and air.
- *Risk aversion and precaution:* Coastal management efforts must adopt a risk-averse and precautionary approach under conditions of uncertainty.
- *Accountability and responsibility:* Coastal management is a shared responsibility. All people must be held responsible for the consequences of their actions, including financial responsibility for negative impacts.
- *Duty of care:* All people and organizations must act with due care to avoid negative impacts on the coastal environment and coastal resources.
- *Integration and participation:* A dedicated, coordinated and integrated coastal management approach must be developed and conducted in a participatory, inclusive and transparent manner.
- *Cooperative governance:* Partnerships between government, the private sector and civil society must be built in order to ensure co-responsibility for coastal management and to empower stakeholders to participate effectively.

Spectrum of marine recreation opportunities model

Mark Orams (1999) suggests that the model developed for the United States Forest Service by Roger Clark and George Stankey (1979) could be used to clarify the diverse range of opportunities for tourists to recreate in the marine environment. These could be viewed as a spectrum of marine recreation opportunities (SMRO) and represented graphically (see Figure 7.1). The spectrum categorizes marine recreation according to its distance from shore because it is this single factor that most strongly influences the activities undertaken, the experiences available and the type of environment in which activities occur. At one extreme of the spectrum, the near-shore environment, tourists are able to undertake a wide variety of shore-based activities which are easily accessible, in an environment

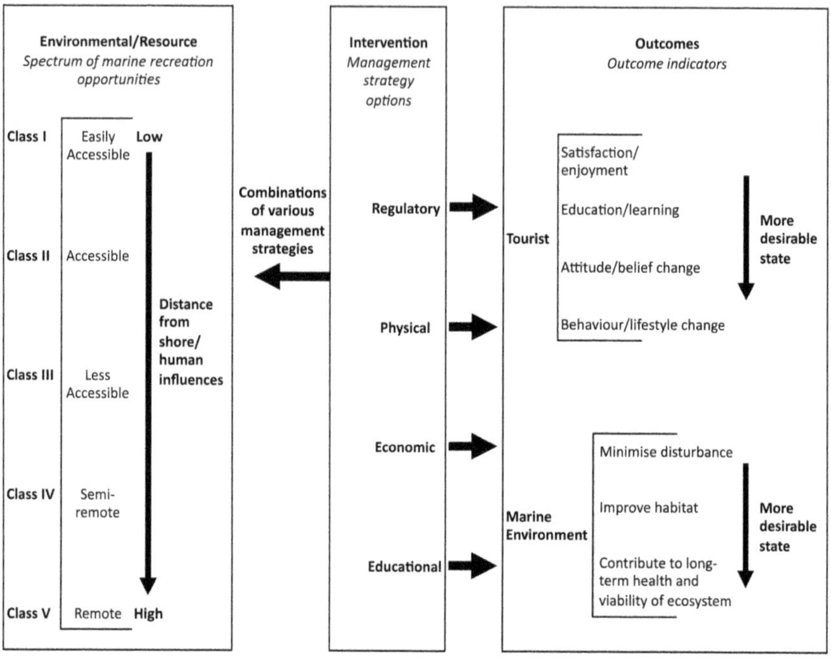

Figure 7.1 SMRO conceptual model (Adapted from Clark and Stankey 1979)

commonly influenced by human-built structures. At the other extreme are those activities that occur far from shore, on the open ocean.

This model, combined with other management techniques and outcome indicators, provides a conceptual framework for considering the management of marine tourism. The overall goal of management strategies designed to control marine tourism is twofold: first, to protect the marine environment from detrimental impacts, and second, to provide for and promote enjoyable tourist experiences. A number of well-established management regimes are utilized to manage tourists and their activities on the sea. These strategies range from no control whatsoever, to complex combinations of structures, technology, economics, regulations and education. To simplify this situation, tourist management strategies can be divided into four main categories: regulatory, physical, economic and educational. Research has indicated that the most effective of the four management strategies is educational. The SMRO conceptual model is illustrated in Figure 7.1.

The sea offers many opportunities for recreation and tourism and in practical terms it is a source of food and transport. In the past, most of our marine environment was 'protected' from tourist use by its inaccessibility, safety concerns and the relatively high cost of recreating in the sea. Through recent decades, significant advances in technology and the increase in international travel have made marine environments more accessible both in real and economic terms. Coastal and marine tourism has become a huge business that forms a significant component of the growing global tourism industry (Orams 1999). However, in many coastal countries, very little research has been undertaken to measure the impact of human recreational activity on coastal and marine tourism environments. An ideal model should incorporate the four tourist management strategies (regulatory, physical, economic and educational), meaning the sustainable future of each development will ultimately depend on the condition of the ocean that connects them.

The role of conservation scientists is crucial in providing answers to the questions of destructive human activity and thus adding to the 'menu' of management strategies available to natural resource managers. It is ironic that whilst almost all of the challenges faced by the natural environment are the result of human activities, including tourism, the great majority of research that occurs on our natural environment remains in the biological and physical sciences. It is therefore critical for our natural environment that conservation scientists incorporate and encourage the social sciences in their community. In particular, there should be a focus on applied research, i.e. that which is focused on providing solutions to problems rather than simply outlining the extent of the problems (Orams 1999, p.97).

Increasing our understanding of humans, what they do and why, is fundamental to finding answers to the challenges posed by marine tourism and other uses of marine resources (Orams 1999). When considering the management of marine tourism activities, the following statement is pertinent:

> So, we have a wide range of management techniques. We can add to them as we realize the possibilities. It is essential that we look at all these possible techniques as a menu, and choose the one, or combination, best situated to the situation.
>
> (Plimmer 1992, p.125)

Orams (1999) suggests that answers to the following questions will provide valuable information for resource managers:

- What motivates people to pursue recreational activities?
- What are the characteristics of marine tourists?
- What techniques are most effective in controlling human behavior in natural settings?
- How can the impacts of recreational activities on natural resources be minimized?
- How can the negative impacts of tourism development on host communities be minimized?
- How can the benefits of marine tourism activities be maximized?
- What techniques are most effective in reducing conflict between competing users?
- What management regimes or decision-making approaches are most effective?

Managing environmental impacts in the marine recreation sector

A Practical Guide to Good Practice: Managing environmental impacts in the marine recreation sector (2009) is the result of a collaborative effort by the Centre for Environmental Leadership in Business (CELB), The Coral Reef Alliance (CORAL) and The Tour Operators' Initiative (TOI). Major contractors of marine recreation services, including tour operators, hotels and cruise lines, are increasingly showing a preference for providers that adopt environmental good practices. Thus, for marine recreation providers, from dive operations to boat rentals to wildlife viewing tours, improving environmental performance can not only contribute to marine conservation and the economic development of coastal communities, it can also improve business, by increasing their attractiveness to companies seeking responsible suppliers. This guide, which is based on an intensive consultation process with industry, local and national authorities, and conservation organizations, is designed for marine recreation providers seeking to understand and adopt environmental good practices.

This guide can also be used as a tool by hoteliers, tour operators and cruise lines in selecting and managing suppliers based on sustainability criteria. For eight key topics associated with marine recreation, the guide offers a brief summary of the potential impacts, the rationale for good practice and suggestions on how to reduce these impacts. Recognizing that an extensive amount of information on environmental good practices for the marine recreation industry already exists, sources for further information are provided at the end of the guide. In the beginning of the guide, a self-assessment checklist has been inserted to allow marine recreation providers to compare their own performance against environmental good practices. This checklist is similar to one that was developed for cruise lines, tour operators and hotels to evaluate their contracted marine recreation suppliers. The questions in the checklist help suppliers understand the issues considered by companies that are trying to make more informed contracting decisions.

Benefits for businesses and communities

Poorly conducted, uninformed or irresponsible marine recreation activities can seriously undermine the health and aesthetics of near-shore marine environments and coral reefs, the very resources that service providers rely on for a healthy business and visitors demand for an enjoyable vacation. Improving environmental and social performance can not only prevent serious impacts, such as loss of tourist revenue, higher unemployment in the tourism sector, lower fish catches and health threats to local people, it can also improve the viability of the tourism industry as a whole. This, in turn, will stimulate economic growth, increase food and employment resources for local populations, and conserve ecosystems such as intact coral reefs that protect coastal communities from waves and storm damage. A conceptual model for an integrated approach to the management of coastal and marine tourism resources is explained with reference to the Broker–Local–Tourist (BLT) model, which is an ideal model for studying the interaction of people and places within any study area.

The Broker–Local–Tourist model

This marine tourism system model observes human and natural settings in a destination and then determines: who are the players in these settings, i.e. brokers, locals and tourists; how the players interact; and what will be the impact of their recreational activities on the social and natural environment. Brokers consist of persons who derive an income or economic benefit from tourism. Locals consist of persons who reside in the general region of a tourism destination but do not derive an income from tourism or engage in its management and regulation. Finally, tourists consist of persons of domestic and international origin who travel for relatively short periods of time for business, recreation and educational purposes before returning home. Tourism and development can be shaped to reflect the best tendencies of humanity. A BLT model of a coastal tourism system is shown in Figure 7.2 (see Miller and Auyong 1991, 1998).

Coastal tourism systems

Coastal tourism systems involve interactions between people and place in destinations that include small communities and villages, self-contained resorts and cosmopolitan cities. These systems change in size and character over time. To understand and ultimately to predict these changes, and also to plan for desired

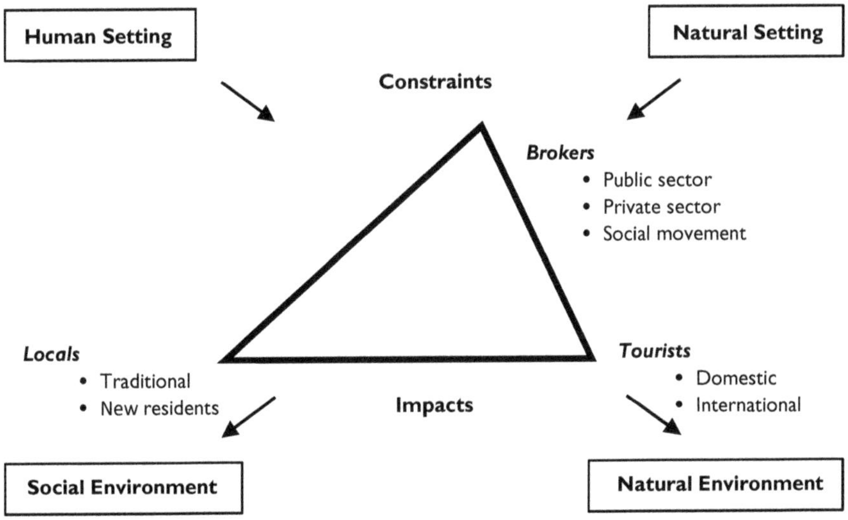

Figure 7.2 Broker–Local–Tourist model (Adapted from Miller and Auyong 1991)

societal and environmental outcomes, analysis must focus on the behavior of components of the system. In this regard, two processes merit attention. First, population dynamics of the BLT model should be monitored. It is not unusual for individuals in the system to change statuses. This can occur as, for example, tourists who visit a coastal destination decide to stay and take on a residence, either as a broker of some kind (for example, as a scuba dive shop entrepreneur or restaurant owner), or as a local (for example, as a lawyer or teacher). Other transformations in status take place as locals change occupations and become private sector brokers by engaging in a tourism business, or become public sector tourism brokers by finding government employment that concerns tourism. Of course, locals and brokers take on the role of a tourist when they vacation on travel of their own.

Second, power dynamics of the BLT model should be taken into account. Tourism is often examined as a product of the aggregate decisions of individual tourists. The relationship forged between the tourist and the local is accordingly depicted as socio-economic in nature; tourists and locals interact as guests and hosts or as consumers and producers. Where power relationships are perceived to exist (as, for example, between First World and Third and Fourth World locals), it is argued that these reveal the colonial and imperialistic leverage tourists have over those whom they visit. From this perspective, tourism systems are controlled and determined often in unfortunate ways by the behaviour of tourists. While there certainly are many instances in which tourists have exercised their influence to selfish and inappropriate advantage in coastal tourism settings, a narrow concentration on the power of tourists can result in analysts missing the power of tourism brokers. As Cheong and Miller (2000) have pointed out, tourists are frequently vulnerable to the power and control of brokers and locals. This is the case when, for example, tourists abide by laws and regulations of public sector brokers, and when they follow the advice and instructions of private sector brokers such as tour guides and travel agents.

Defining coastal and marine tourism destinations

The following text has been extracted and condensed from a paper presented by Professor Marc Miller at the 7th International Coastal and Marine Congress convened at Breda University, the Netherlands, in 2012 (Miller 2012), in which Miller argues the definition of 'destinations'.

Introduction

Coastal and marine tourism destinations, as with tourism destinations elsewhere, reveal the contradictions of travel. In the first place, touristic destinations are Janus-faced in having both real and unreal aspects. Destinations are real and concrete in that they are actual places that can be objectively described, measured, owned and governed. They are unreal and ethereal in that they also exist in our dreams and hopes. Coastal and marine destinations are found at many latitudes, and come in an astonishing array of sizes and shapes. When asked about their destination, tourists respond by referencing specific continents, coastlines, water bodies, reefs and trenches, mountains, routes, nations, provinces, cities, towns, communities, resorts, theme parks, festivals and events of many types, historical sites, and more.

Tourists travel to see particular (sub-) cultural, ethnic, political and religious groups of people. They also travel to see special edifices and technologies (e.g. buildings, stadiums, churches, bridges, monuments, ships, waterfronts, etc). Sometimes tourists answer the destination query by naming a moving attraction, as when they indicate they want to 'see whales' or 'see migrating Arctic terns'. When tourists are not exactly sure where they will end up, and it is common for some tourists not to want to know, they identify unnamed destinations such as 'the wilderness' or 'the tropics'. With branding, destinations can quickly change their boundaries; and destinations also morph with the creation and deletion of amenities.

In this process, and in response to the pressures of disinvestment and the quantity and type of tourists attracted, destinations can be transformed into entirely new kinds of places. With personal inspection of a destination, vision comes face-to-face with reality. It is at this time that expectations are tested; the results take shape as we feel elation or disappointment. We then adjust or abandon our destination expectations. In the second place, destinations reveal travel contradictions as they take on meaning for us in accordance with our personal values. It is important to understand that the meaning of a destination is influenced by one's position in a tourism system. For tourists, destinations are generally understood as places they want to visit although there certainly are destinations tourists want to avoid. For tourism brokers destinations are places which they are involved in deliberately building, protecting, restoring and marketing. Locals are in the sometimes uncomfortable position of witnessing the interplay of the destination demand created by tourists and the destination supply provided by brokers. What tourists and brokers relate to as destinations, locals call 'home' – and the quality of home can be affected both positively and negatively by destination planning, design, management and visitation.

Tourism destinations: Varying definitions

To preface a discussion on the nature of coastal and marine tourism destinations, two contemporary definitions should be considered. First, the United Nations World Tourism Organization, an intergovernmental entity responsible for the promotion of responsible, sustainable and universally accessible tourism, provides this definition:

> Destination is a significant place visited on a trip. It may be defined as the farthest point away from home visited (distance destination), the place

> where the most amount of time was spent (main destination), or place the visitor thinks of as the primary place visited (motivating destination).
>
> (UNWTO 1995, p.7)

Second, *The Encyclopedia of Tourism and Recreation in Marine Environments* defines destination as follows:

> Destination: In general terms, a destination is any place that a person travels to, for whatever reason. More specifically, a tourism destination is a place that attracts visitors for a temporary stay, and features attractions, amenities, accommodation and a host community. Destinations range in size from purpose-built resorts to villages, cities, provinces, states, countries, and continents. Destinations of all sizes often cluster with neighbouring areas to form macro-destinations. In marine environments, destinations range in scale from small beaches and islands to relatively large stretches of coastline and trans-national regions.
>
> (Lück 2008)

For decades, tourism experts in academia, research institutes and think tanks, and in consulting and other private practice have struggled to provide a lasting, precise and pragmatic definition of 'destination' as a unit of analysis. In this quest, they have resembled a host of social scientists (e.g. cultural anthropologists, sociologists, geographers, economists and political scientists), urban planners and landscape architects, historians, and other professionals (e.g. in the fields of business and public affairs) who in similar fashion have aspired to generate useful definitions of such terms as (sub-) culture, community, place, environment, landscape, historical era, etc, among other foundational concepts in their fields. To plan, design and manage coastal and marine destinations in a pragmatic and responsible manner, it is incumbent on tourism practitioners to establish a set of rather standard definitions and to operationalize tourism concepts in practical ways. Destinations have different kinds of significances for different kinds of people (e.g. tourists, analysts, policymakers or entrepreneurs).

The meaning of destination hinges on the purposes for which it is defined in the first place. Thus, any destination definition only has value to the extent to which it helps people to achieve their particular ends. In his paper, Miller (2012) provides a sample of how the notion of a tourism destination has been conceptualized and operationalized. Different types of tourist destinations compete for the same markets: for example, centrally managed destinations owned by one or few companies or individuals, destinations consisting of a system of fragmented services delivered by many companies and actors, and traditional tourist regions/destinations with community character. Destinations are the competitive units of incoming tourism. However, most would agree that destinations are therefore an important part of a tourism product.

A destination development perspective

Richard Butler (1980), a pioneer tourism researcher, has typified the process by which destinations have natural histories, suggesting that individual destinations experience the following stages (see Figure 7.3):

- *Exploration:* A small number of tourists visit the area. The area is unspoilt and few tourist facilities exist.
- *Involvement:* Local people start to provide some facilities for tourists. There starts to become a recognized tourist season.
- *Development:* The host country starts to develop and advertise the area. The area becomes recognized as a tourist destination.
- *Consolidation:* The area continues to attract tourists. The growth in tourist numbers may not be a fast as before. Some tensions develop between the host and the tourists.
- *Stagnation:* The facilities for the tourists may decline as they become old and run down. The numbers of tourists may decline too.

Then, depending on constraints and pressures bearing on the situation, the destination will experience one of the following stages:

- *Rejuvenation:* Investment and modernization may occur which leads to improvements and visitor numbers may increase again.
- *Decline:* If the resort is not rejuvenated, then it will go into decline. People lose their tourism-related jobs, and the image of the area suffers.

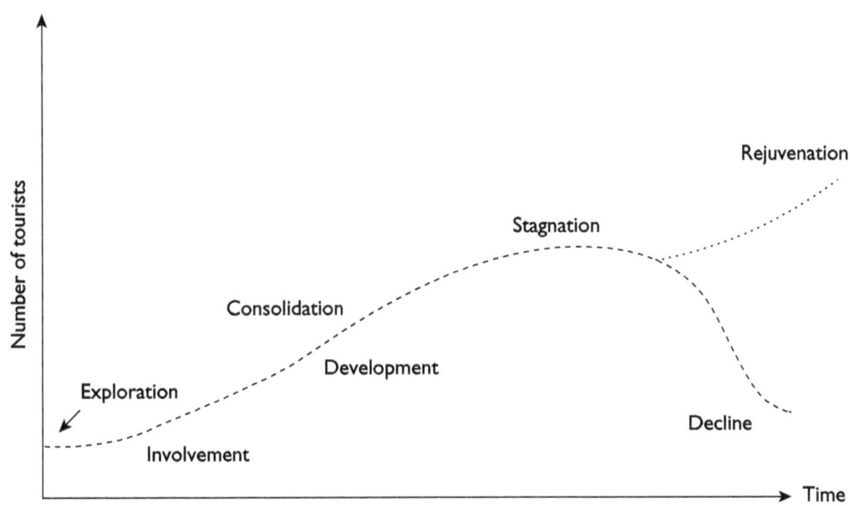

Figure 7.3 Butler conceptual model (Adapted from Butler 1980, 2011)

Butler (2011) has suggested the following definition:

> Tourist Destinations: A place where tourists plan to spend time away from home. This geographical area could be as small as a self-contained centre, such as a village, town or city, or be as broad as a region, island or country. It encompasses all the organizations, companies, individuals and government bodies (stakeholders) which offer products and services to people visiting the destination, as well as natural and artificial resources and attractions. More than one destination may be visited as part of a holiday, for example, on a tour or cruise. The term is considered complex to define but it tends to refer only to the geographical unit in which products can be purchased or experienced.

A destination image perspective

It is a truism that coastal tourism destinations flourish or fade in large part due to the effect of the destinations on the decision making of the tourist. Temporally, tourists have a pre-trip image of a destination rooted in their personal experience there (if any) and information (some valid, some misleading, some wrong) gleaned from tourism marketers, travel critics, news accounts, friends, family and other sources. This image directly influences tourists' selection of destinations to visit, and the standards they use to subjectively measure their satisfaction at destinations visited. Attributes of tourism destinations and attractions which help to shape the behavior and motivation of tourists pertain to: natural resources, cultural resources, recreational facilities, events, specific activities and amenities, and the psychological appeal (e.g. romance, adventure, remoteness of the destination, etc).

A landscape perspective

Landscape architecture and regional planning professionals use the term 'landscape' to encompass systems which have human and natural markers. 'Land' means both a place and the people living there. The language of landscape recovers the dynamic connection between place and those who dwell there. Humans are not the sole authors of landscape. Mountains, gardens and cities are shaped by volcanoes and rain, plants and animals, and human hands and minds. 'Landscapes are as small as a garden, as large as a planet': with this usage of the term, all destinations may be treated as landscapes. Landscape professionals are as committed to the spaces between pristine nature and cities as they are to towns and cities *per se*. Positive valuations of landscapes with human, natural and technological features are also increasingly held by conservationists and natural scientists.

CASE STUDY: SOUTH AFRICA, BULUNGULA LODGE COASTAL COMMUNITY PARTNERSHIP

Introduction

Former President Nelson Mandela made the following statement:

> Africa's long and beautiful coasts and the abundance of marine resources can contribute to providing economic, food and environmental security for the continent. These coastal and marine resources, like the rest of Africa's environmental resources, cannot continue to be exploited in a manner that does not benefit Africa and her people. This is a paradox of a people dying from hunger, starvation and poverty when they are potentially so rich and well endowed.
>
> (Mandela, Cape Town, 1998)

Bulungula Lodge is living proof that such an ideal of 'benefitting Africa' can be achieved. Bulungula was one of the first lodges in the world to be fair trade accredited; the community owns 40% of it. David Martin, the private sector investor who initiated the joint venture partnership, has made it his priority to draw on local resources in ways that enrich the community.

Guests can sign up for various activities, all of which are operated by locals independently of the lodge. Activities include canoeing, fishing, horseback riding, a woman power tour, and a tour through the forest with a natural herbalist. It has a rugged feel, and is therefore refreshing compared to the typical structured tours, such as those which attempt to give insight to the authentic way of life in South Africa. In addition to the preservation of the community and the many ways individuals in the community are empowering themselves through Bulungula Lodge, the lodge is incredibly conscious of preserving natural resources and the environment. The few appliances and lights are run on solar power. In the decorated mud hut lodges, two candles and a box of matches are provided for guests. The toilets are odourless pit latrines. The rocket showers run on a small amount of paraffin (kerosene) poured into the base of the pipe that serves to heat the water.

The food is delicious, sometimes a local dish, sometimes a taste of the exotic. Every evening, a vegetarian version of the dinner is available. Anytime of the day or night guests can help themselves to drinks or snacks available without having to pay cash up front. Bulungula Lodge is an outstanding community–private partnership model that reflects passion, preservation, culture, empowerment and beauty.

Background

The Wild Coast of South Africa (see Figure 7.4) has long been regarded as one of the world's most beautiful coastal wildernesses. For decades experts have said that it is just a matter of time before international tourism discovers the endless beaches, stunning views and culturally rich communities of the Wild Coast. Over the last ten years this coast has benefited from the enormous growth of tourism in South Africa, but until recently the benefits were mostly limited to a handful of backpacker hostels in Coffee Bay and Port St Johns. The poor infrastructure and a lack of investors left much of this area's natural wealth untapped to the detriment of the local communities and the environment. The fact that most tourism infrastructure has been situated mainly in two towns has resulted in the anomaly that there are virtually no lodges close to the beach, and thus the classic beach-paradise lodge, so popular in Thailand and Malawi, was not to be found anywhere along the Wild Coast's 300 km of beach until fairly recently.

The area surrounding the Bulungula river mouth is one of the most beautiful on the entire coast. Lush coastal forest, a clear blue river and a long white beach meet at the foot of rolling hills dotted with the huts of Nqileni village. Unfortunately, the lack of a road has meant that previously the site's obvious tourism potential had not been exploited and thus the local community eked out a subsistence lifestyle without any prospect of gainful employment. By way of an example, a visit to the nearest clinic required a wheelbarrow ride of 4 km, a boat ride across the Xora River, another wheelbarrow ride up the hill, and then half a day's wait for a taxi which may never arrive.

The Bulungula Lodge community-based tourism project began in October 2002 when a local entrepreneur, Dave Martin, identified the site for a possible joint venture between the Nqileni community and himself. The lodge was built with capital of R800,000 and has created 20 permanent jobs in a community where there was previously 100% unemployment. One job in a rural community will often support at least ten members of the extended family at a subsistence level. The lodge has also facilitated the creation of 13 community-owned tourism businesses that provide services ranging from massages to fresh vegetables to forest tours with a herbalist. In total these small businesses have created another 23 jobs, bringing the total employment generated by this project to 43 long-term jobs. In addition to the community's share, lodge profits have been used to build a classroom at the local school.

The culture of Bulungula Lodge is an open one. There are no fences, keys or locks at Bulungula. Community members are welcome in the restaurant and bar at any time of the day or night and it is this cooperative environment that has resulted in crime being nonexistent. The unique atmosphere

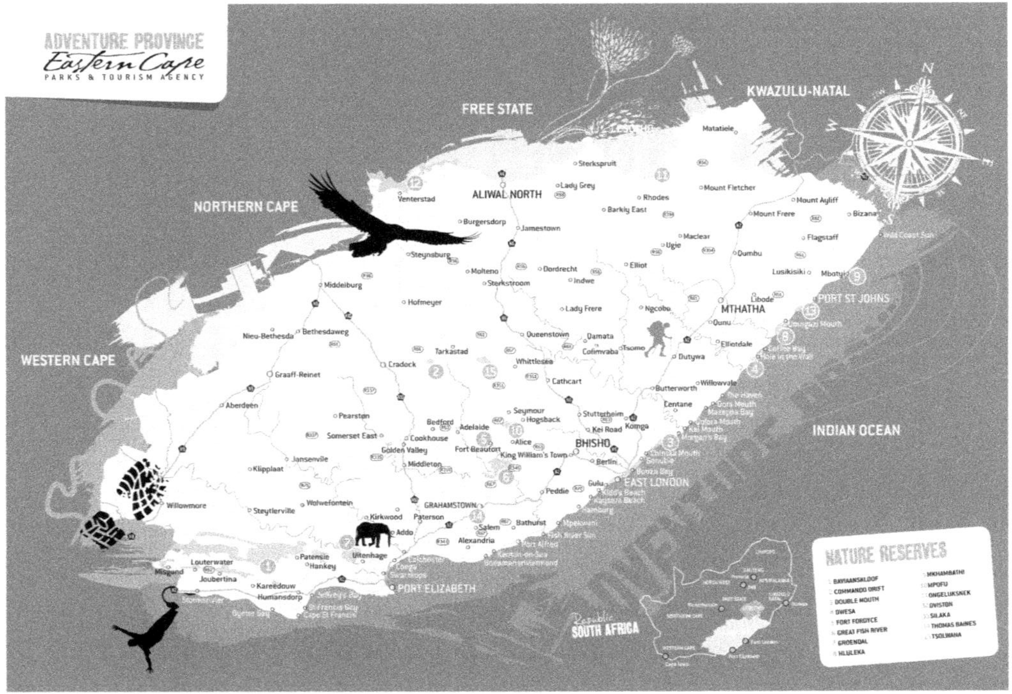

Figure 7.4 Map of the Wild Coast, Eastern Cape, South Africa

(*Source:* Eastern Cape Parks & Tourism Agency, with permission.)

has resulted in Bulungula receiving many accolades, including being named by the Rough Guide as 'One of the World's Top 25 Ultimate Ethical Experiences' in 2007, as well as being named 'Second Best Eco-lodge in the World' by Lonely Planet and *The Guardian* in 2014.

Most guests initially plan to stay for a few days but once they have experienced this pristine coastal environment, and the hospitality of the local communities, they inevitably stay longer. Bulungula Lodge is a benchmark for appropriate coastal community-based development in a rural environment.

Wild Coast development policy

The Wild Coast Development Policy provides spatial guidelines for appropriate tourism development along the Wild Coast. The following controls on tourism developments aim to ensure that an appropriate type and form of tourism development is encouraged at the appropriate location, and they will also ensure that developments are complementary:

- *First order nodes:* The most urban and extensively developed areas, of a 'seaside resort' nature, such as Port St Johns or Coffee Bay. Within these areas, the main focus should be on recreation provided by the development, not the environment, although the development is located in a pleasant, clean and attractive setting. These nodes are suitable for large hotel and cluster developments, with some cottage development.
- *Second order nodes:* Less developed and urban in nature, being more focused on 'family holiday' tourism and recreation facilities, provided by both the development and the environment. Within these nodes, cottages, cluster complexes and family hotels may be developed.
- *Ecotourism/low impact tourism zones:* Zones within which ecotourism (defined for the purpose of the policy as natural-resource based, low intensity, environmentally and culturally sensitive tourism) developments and activities are emphasized.

Tourism development, with the exception of ecotourism/low impact tourism development, should be concentrated in the identified discrete first and second order nodes. There should be no tourism development within those areas which are not delineated as first order nodes, second order nodes or ecotourism zones. Within each node, careful attention should be given to ensuring that there is a compatible mix of tourism and recreational uses, in order to avoid conflict between land uses and activities. The location of tourism development along the Wild Coast should also be consistent with the Integrated Development Plans or Local Development Plans prepared for the region. Additional controls on development may be warranted by the environmental characteristics of the area.

Bulungula Lodge is located in an ecotourism/low impact tourism zone. There is a need in the Wild Coast region of South Africa for a successful tourism development model that can be replicated in areas identified as priority areas for sustainable tourism development. This model should take into account the key issues of land usage, land ownership, partnerships and self-employment. If there is one thing that local people living in the area value more than anything else it is a sense of land ownership. The future of sustainable tourism development in the Wild Coast region will therefore depend on projects that generate self-employment and small business opportunities rather than employment dependent upon large-scale resort and hotel chain development.

The goal of sustainable tourism development in the Wild Coast region should incorporate a management program that will enhance the positive contributions of tourism development along the Wild Coast while minimizing negative consequences in the affected towns and villages, with specific reference to local communities and the pristine environment. Development is defined here as directed change resulting in greater empowerment of those who are affected by that change. To encourage successful development along the Wild Coast, a community approach should be used to involve the affected communities in the decision-making process from the planning stage through project evaluation. A community approach focuses on how tourism must be understood within the larger context of social and economic conditions (Murphy 1985).

In determining the conservation estate (i.e. just how much land should be conserved, who owns the land and how should it be managed), a number of historical factors need to be taken into account, especially where issues of land ownership are involved. Issues relating to current land tenure are extremely complex and emotive but are critical to the success of any proposed development in the region. In 2006, the Cabinet approved an Integrated Coastal Management Bill to protect the country's coastline. This Bill provides additional legal teeth to deal with free-for-all land grabs and developments. A study of the historical background to rural coastal development in South Africa is important to understand the context of the Bulungula Lodge development, which in every way adheres to all the tourism and environmental guidelines outlined in the Wild Coast Tourism Development Policy (17 July 2000).

Bulungula has complied with the highest standards of global responsible, green, ethical and sustainable tourism best practice. In this context one might ask, what are the prospects for sustainable development on the Wild

Coast? The Wild Coast clearly needs urgent intervention to address the appalling poverty in the region, and effective service delivery should be made an urgent priority. However, as concerns the creation of sustainable jobs, the government should create an enabling environment for the private sector to create wealth. When looking at the Wild Coast there are two industries that stand out as having potential for sustainable development: agriculture and tourism. David Martin believes that the Bulungula Lodge case study will give insight into the reasons tourism has apparently failed to deliver. To understand why the Wild Coast is not teeming with similar community-based lodges, it is necessary to understand the tortuous process required to have the lodge approved.

In December 2002 the community of Nqileni village and David Martin discussed and agreed on the establishment of a community-based eco-lodge as a joint venture. A proposal was produced and presented, using the Eastern Cape government's Wild Coast Tourism Development Policy as a blueprint and working in partnership with the provincial departments of environmental affairs and land affairs. After completing the necessary environmental report, permission was swiftly obtained from the department of environmental affairs. The difficult part was getting the necessary lease issued by the department of land affairs. As a lease had at that stage not been issued on the Wild Coast there was no precedent to follow. This resulted in a complicated process which invariably resulted in officials trying to pass the responsibility upwards to officials with greater authority. Finally, in March 2004, an interim lease (the first apparently) was granted, which was to be used during the building of Bulungula Lodge while they waited for the long lease (30 years) to be signed by the national minister of land affairs within six months. A sensible approval process must be created in line with the Wild Coast Tourism Development Policy that has realistic and acceptable timescales. This office will need to be well resourced with skills and money to assess correctly and speedily the viability of proposed projects. In addition this office should proactively identify and finalize leases for, say, ten sites prior to inviting investors to tender bids to be the private-sector partner for the local communities.

Assessment of performance against the three pillars of sustainable tourism

A UN report (WCED 1987) defined sustainable development as:

> Development that meets the needs of the present without compromising the ability of future generations to meet their own needs.

A thorough assessment undertaken by Fair Trade in Tourism South Africa (FTTSA) in 2006 confirmed that Bulungula satisfied the requirements for

FTTSA certification. Fair Trade in Tourism is about ensuring that the people whose land, natural resources, labour, knowledge and culture are used for tourism activities, actually benefit from tourism. The criteria is based on fair wages and working conditions; fair operations, purchasing and distribution of benefits; ethical business practice; and respect for human rights, culture and the environment. The findings of the assessment are now discussed, as relating to economic, environmental and social performance.

Economic performance

The lodge is contributing towards poverty alleviation in the community and some of the economic benefits to the local community include:

- Employment opportunities; income from the lease agreement; income and employment from the 100% community owned and run tourism businesses (horse-riding, canoeing, fishing, guiding, baking, sewing, cooking and wood-carving have created jobs and income for an additional 33 families); and profits paid to the community trust as part owners.
- All employees are selected by the local community in a community meeting. Non-community members are only eligible for jobs that cannot be filled by local people due to a lack of skills.
- Current growth plans focus on increasing the number of community tourism businesses, which will provide more entertainment for guests and thus keep guests for longer stays and benefit the community through increased income via the community businesses.
- The lodge is 40% owned by the community and declared a dividend of R20,000 in February 2006. It is on target to declare larger profits in the future. There is scope for more tourism infrastructure development based on demand and supply without compromising the environmental integrity of this pristine area.

Environmental performance

Bulungula believes very sincerely that they are a part of the solution to the planet's environmental problems. Although they are not pushy environmentalists, they are systematically developing the lodge as an example of how one can live in harmony with the environment through reduced consumption, appropriate technologies and creative thinking. There is still room for improvement, but they are constantly trying to reduce their carbon footprint.

Bulungula's electricity is supplied by solar panels (on the roof of the first store room). These panels are used to run the lights, music, satellite phone, computers, water pumps and various battery chargers. They have a large battery bank that can supply the lodge with electricity for up to 7 days of cloudy weather. The key with solar energy is to be constantly aware of how much electricity is being used and to ensure that all appliances are as energy efficient as possible. The entire lodge uses in 24 hours the same amount of energy as a toaster or a kettle would use in 1 hour.

The stoves and fridges run on natural gas which is the cleanest of the conventional fossil fuels and the showers heat water with 100 ml of paraffin. The hope is to introduce biofuel soon. The cost savings this approach has achieved is phenomenal. A large backpacker lodge on its own can easily spend R12,000/month on electricity, while Bulungula spends around R4,000/month for all its energy needs.

Bulungula's water comes from two sources: rainwater for laundry and underground water pumped via solar panels for drinking, washing and everything else. Again, the aim is to reduce consumption. In this regard, the use of compost toilets and water-conserving showers ensure that about 40 litres of water is used per guest per night (whereas the average for developed-world consumption is about 250 litres/person/day). Flush toilets are one of the world's most environmentally destructive inventions (i.e. human excretion is flushed directly into 40% of the world's piped drinking water): at Bulungula, water is cycled through a banana circle behind the showers. All shower and basin water goes into the centre of the banana circle, creating a wet sponge. The banana and papaya trees send their roots down into this wet nutritious sponge and grow big extremely quickly, thus converting dirty shower water into healthy, juicy bananas, papayas and various other vegetables. Outside the kitchen window is a three-pond filtration system which removes oils, fats and food particles from the dirty dishwater and feeds this water into another small banana circle.

Paper and cardboard is burned at Bulungula and all the rest of the rubbish is taken to Mthatha. In 2009, primary dune forest was fenced off as goats and other domestic animals were decimating it. Goats eat all the undergrowth including the young trees which means that when the old trees fall, there are no new trees to replace them and the forest dies. This lack of undergrowth allows invasive alien plants like Lantana (pretty multicoloured flowers) and Inkberry (light-green, leafy bush) to take over as they are poisonous and so are not eaten by the goats. Since putting up this fence and removing all the invasive plants, the forest has recovered amazingly. By comparison, all that can be seen in the forest immediately south of this fence is old trees and beach sand beneath them.

Finally, over 10,000 mangrove seeds have been planted around the stumps of the dead mangroves in the river. These mangroves died during the drought of 1997. Many seeds have taken root and there are now over

100 trees of five-plus years old – at Bulungula, they are hopeful that the mangrove forest will thrive again.

Social performance

Bulungula Lodge is making very positive attempts at addressing the issues of health, education and infrastructure in the village. The lodge is investing time, effort and money in these areas to help the community to realize their human, social and economic potential in the following ways:

- The Nqileni community appreciates the presence of visitors in their village and is encouraged to interact with the tourists. The community has developed a sense of pride within their village as a result of tourists from all around the world being impressed during their stay. The local people manage all the local excursions and activities.
- The lodge has interacted with the local municipality regarding the need for improved infrastructure in the community, as well as with a few of the schools and hospitals in the area. Doctors, nurses and local teachers visit the lodge. Schools have held 'teacher appreciation days' at the lodge and the hospital has used the lodge as a venue for training programs.
- HIV education and free testing days are done regularly for staff.
- The lodge provides an ambulance service to transport the ill to clinics or hospital.
- Temporary English language classes are offered to help the community learn English, and thus have an improved ability and confidence to interact with tourists.
- The lodge has built toilets in the community and a guest has donated improved cooking stoves.
- A collection of Xhosa children's books has been gathered for the library.
- Lodge staff ensure that, on arrival, customers are well briefed and informed of the cultural norms and socio-economic conditions within the community, as well as generating awareness on how the lodge is practicing sustainable management. This adds value to the tourists' experience and makes them feel involved in contributing to poverty alleviation and job creation.
- By encouraging social encounters with local people, the lodge allows the tourist to get closer to local culture. The lodge is open to community members to socialize with guests informally and there are basic icebreaker phrases printed on the toilet doors for the tourists to try out!
- The lodge provides clean drinking water for the surrounding village.

- Using solar power, the lodge charges community cell phones for free.
- The lodge equips the local community with free basic first aid and medicines.

Conclusion

Bulungula Lodge is committed to fulfilling its role as a sustainable development initiative for the local community and is actively incorporating the three pillars of sustainability into its day-to-day business activities. It uses the local people, their land, culture, heritage and social fabric as the 'ingredients' for its product and works together in a close partnership with the community to ensure ethical business practices that will benefit its stakeholder groups.

Bulungula demonstrates strategic intervention by a private sector entrepreneur as opposed to strategic intervention by government and its local economic development agencies. Successful tourism development along the Wild Coast requires the right partnerships, i.e. an entrepreneur with the skills to develop sustainable tourism and a responsive host community willing to cooperate in a joint venture. This requires building a relationship of trust and mutual respect.

The lodge has clearly demonstrated that it not only creates employment but also provides other benefits like a free ambulance service. Community self-image has been substantially improved as thousands of visitors call their village paradise. Visitors have donated hundreds of thousands of rand for various community projects. In December 2014, just 7 years after the first tourists arrived in the village of Nqileni, David Martin has transferred ownership of the lodge from 40% share to 100% wholly owned by the local community.

Discussion

Throughout this book, different kinds of mutually beneficial cooperative alliances are being examined – regardless of whether they are maritime clusters or informal alliances or partnerships. Successful partnerships, if they are community based, require a willing entrepreneur and a willing local community. Perhaps the key to the Bulungula success story is that the project has been driven by a local entrepreneur with a passion for the country of his birth. Furthermore, his incredible local knowledge and sensitivity to the needs of local pro-poor communities are matched by his commitment to a joint venture partnership that will deliver benefits to the host community. However, equally as important is that the Nqileni community was willing to enter into an agreement that was not too demanding in terms of the investment capital required to build the infrastructure and also in terms of the length of tenure to allow the investor sufficient time to recover his/her capital plus interest on any loan capital.

References and further reading

Butler, R.W. (1980). 'The concept of the tourist area life-cycle of evolution: Implications for management of resources'. *Canadian Geographer*, 24(1): 5–12.

Butler, R.W. (2011). *Tourism Area Life Cycle*. Oxford: Goodfellow Publishers.

CCMAP (2012). 'Integrated coastal zone management: Lessons in capacity building and good governance for coastal adaptation'. CCMAP Issue Brief, Series No.1. www.start.org/download/publications/ccmap1-coast-No%201.pdf (accessed: April 2017).

Celliers, L., Breetzke, T., Moore, L. and Malan, D. (2009). *A User-Friendly Guide to the Integrated Coastal Management Act of South Africa*. Cape Town: The Department of Environmental Affairs and SSI Engineers and Environmental Consultants.

Centre for Environmental Leadership in Business (CELB), The Coral Reef Alliance (CORAL) and The Tour Operators' Initiative (TOI) (2009). 'A practical guide to good practice: Managing environmental impacts in the marine recreation sector'. http://destinet.eu/resources/...-various-target-groups/individual-puplications/A_practical_guide_to_good_practice_Managing_environmental_impacts_in_the_marine_recreation_sector.pdf (accessed: April 2017).

Cheong, S-M. and Miller, M.L. (2000). 'Power and tourism: A Foucauldian observation'. *Annals of Tourism Research*, 27(2): 371–390.

Clark, R.N. and Stankey, G.H. (1979). 'The recreation opportunity spectrum: A framework for planning, mangement, and research.' General technical report. https://iucn.oscar.ncsu.edu/mediawiki/images/b/b4/Clark(1979).pdf (accessed: May 2017).

Coastal Zone Management Authority and Institute (CZMAI) (2016). *Belize Integrated Coastal Zone Management Plan*. ('Promoting the wise, planned use of Belize's coastal resources'.) Belize City: CZMAI.

Department of Economic Development, Environmental Affairs and Tourism (DEDEAT) (2013). *Environmental Management Plan for the Wild Coast*. Eastern Cape Province: DEDEAT.

Department of Environmental Affairs and Tourism (2006). 'National environmental management: Integrated coastal management bill'. www.environment.gov.za/sites/default/files/legislations/nem_integratedcoastal_management40.pdf (accessed: April 2017).

GESAMP (Group of Experts on the Scientific Aspects of Marine Environmental Protection) (1996). 'The contributions of science to integrated coastal management, reports and studies'. Reports and Studies No.61. www.gesamp.org/publications/publicationdisplaypages/rs61 (accessed: April 2017).

Haines-Young, R. and Potschin, M. (2011). 'Integrated coastal zone management and the ecosystem approach'. CEM Working Paper No.7. Nottingham: Center for Environmental Management.

Hoegh-Guldberg, O. (2015). *Reviving the Ocean Economy: The case for action – 2015*. Geneva: WWF International.

Ibrahim, H.S. and Shaw, D. (2012). 'Assessing progress toward integrated coastal zone management: Some lessons from Egypt'. *Ocean & Coastal Management*, 58: 26–35.

Le Tissier, M. D'A.A. and Hills, J.M. (2010). 'Practitioner training for building capacity in ICZM'. *Ocean & Coastal Management*, 53: 787–795.

Lück, M. (ed) (2008). *The Encyclopedia of Tourism and Recreation in Marine Environments*. Oxfordshire, UK: CABI (Centre for Agriculture and Biosciences International).

Mee, L. (2012). 'Between the devil and the deep blue sea: The coastal zone in an era of globalization'. *Estuarine, Coastal and Shelf Science*, 96: 1–8.

Miller, M.L. (2012). 'Editorial' and 'Defining coastal and marine tourism destinations' in: Marc L. Miller, J. Auyong, M. Smeenge, B. Offringa, et al., *A Field Focuses on Destinations: Papers from the 7th International Coastal and Marine Congress*. Putnam Valley, NY: Cognizant Communication Corporation.

Miller, M.L. and Auyong, J. (1991). 'Coastal zone tourism: A potent force affecting environment and society'. *Marine Policy*, March: 75–99.

Miller, M.L. and Auyong, J. (eds) (1998). *Proceedings of the 1996 World Congress on Coastal and Marine Tourism: Experiences in management and development*. Washington, D.C.: Washington Sea Grant Program and the School of Maritime Affairs, University of Washington; and Oregon: Oregon Sea Grant College Program, Oregon State University.

Moksness, E., Dahl, E. and Stottrup, J. (eds) (2013). *Global Challenges in Integrated Coastal Zone Management*. Chichester and New York: Wiley-Blackwell.

Murphy, P.E. (1985). *Tourism: A community approach*. London: Methuen.

Orams, M. (1999). *Marine Tourism: Development, impacts and management*. London and New York: Routledge.

Organisation for Economic Co-operation and Development (OECD) (2016). *The Ocean Economy in 2030*. Paris: OECD Publishing.

Pickaver, A., Steijn, R., Ferreira, M., Czerniak, P., et al. (2011). *Integrated Coastal Zone Management: Participation practices in Europe*. Luxembourg: Publications Office of the European Union.

Plimmer, W.N. (1992). 'Managing for growth: Regulation versus the market' in: *Proceedings of Conference Ecotourism Business of the Pacific*. Auckland: University of Auckland, p.122–125.

Province of the Eastern Cape (2014). 'National Environmental Management Act (107/1998): Wild Coast Environmental Management Plan'. Provincial Gazette Extraordinary, No.3210. www.wildcoast.co.za/files/Government-Gazette-Wild-Coast-EMP-03July2014.pdf (accessed: April 2017).

Schindler, S. (2006). 'Developing sustainable tourism in a Wild Coast community – The case of Bulungula Lodge'. UNISA Centre for Corporate Citizenship and Fair Trade in Tourism South Africa (FTTSA).

Tabet, L. and Fanning, L. (2012). 'Integrated coastal zone management under authoritarian rule: An evaluation framework of coastal governance in Egypt'. *Ocean & Coastal Management*, 61: 1–9.

UNEP (United Nations Environment Programme) (1995). 'The global program of action for the protection of the marine environment from land-based activities'. https://wedocs.unep.org/bitstream/handle/20.500.11822/11144/wbrs18_pre_%20%288%29.pdf?sequence=1&isAllowed=y (accessed: May 2017).

United Nations (2014). *UN Sustainable Development Goals – Oceans*. 16th Global Meeting of the Regional Seas Conventions and Action Plans: Athens, Greece.

United Nations Economic Commission for Africa (UNECA) (2016). *Africa's Blue Economy: A policy handbook*. Addis Ababa, Ethiopia: UNECA.

UNWTO (World Tourism Organization) (1995). *Concepts, Definitions and Classifications for Tourism Statistics*. Madrid: UNTWO.

WCED (United Nations World Commission on Environment and Development) (1987). 'Report of the World Commission on Environment and Development: Our common future'. www.un-documents.net/our-common-future.pdf (accessed: May 2017).

Chapter 8

Small island developing states

Introduction

A European maritime cluster might be defined as an organization that aims to promote and enhance the strength of maritime sectors and activities. Indeed, maritime clusters are useful platforms for the involvement of all stakeholders, and positive proof of this can be found in European countries that have made a success out of maritime clusters, such as France, Germany, Holland and Belgium. When assessing the local scenario from this European perspective, small island developing states (SIDS) are somewhat lagging behind. The island mentality makes local stakeholders introspective and lacking a view of the wider picture.

The fear of sharing information with competitors hinders innovation and development. Furthermore, there is too much fragmentation within this industry which, within a European context, is too small to take any advantage emanating from economies of scale. Paradoxically, it is due to this reality that small islands need to cluster together all the segments of the maritime industry. The small size of islands, and relatively limited turnover at industry level, are at times precluding them from participating in EU tenders. Taking the cue from European success stories in maritime clustering, the French maritime cluster is comprised of 42 cross-sector groupings.

SIDS are a distinct group of developing countries that face common social, economic and environmental challenges. These include small populations, high dependency on development assistance and international trade (especially commodities through preferential trade regimes), susceptibility to external shocks, high transportation costs and low connectivity, susceptibility to natural disasters, and high vulnerability to the impacts of climate change. Of particular concern to SIDS are the risks associated with rising ambient temperatures and sea levels.

While small islands share common characteristics, they are not homogenous and have diverse social and economic structures. These structures may well define the policy approaches pursued and the extent to which additional opportunities from the oceans can be harvested. The level of dependence on oceans by SIDS differs from country to country. For example, some are reliant on natural resources to drive economic development due to factor endowments. Resource-dependent

SIDS include Nauru, Papua New Guinea, and Trinidad and Tobago, which rely heavily on either oil, gas, phosphates, timber or fish exports. Other SIDS – including Barbados, Mauritius, Saint Lucia and the Seychelles – are more services-oriented, with an emphasis on tourism and some financial services. For SIDS, oceans and seas constitute a much larger geographic area than their inland territory, especially when the exclusive economic zone (EEZ) is taken into account. The Cook Islands, for example, have a land space of 240 square kilometres and an EEZ of 1.8 million square kilometres.

Although they are afflicted by economic difficulties and confronted by development imperatives similar to those of developing countries generally, SIDS have their own peculiar vulnerabilities and characteristics. Their small size, remoteness, narrow resource and export base, external economic shocks, and exposure to global environmental challenges (including a large range of impacts from climate change, and potentially more frequent and intense natural disasters). SIDS could significantly benefit from a collaborative economic development approach and in this regard maritime clusters are the perfect vehicle for facilitating such collaboration.

Policy frameworks for SIDS

The Third International Conference on SIDS was held in Apia, Samoa, in September 2014, with the overarching theme of 'The sustainable development of small island developing states through genuine and durable partnerships'. Nearly 300 partnerships were announced at the conference and monitored through the Partnership Platform. The SIDS Accelerated Modalities of Action (SAMOA) Pathway (Samoa Pathway) adopted at the Conference addresses priority areas for SIDS and calls for urgent actions and support for SIDS' efforts to achieve their sustainable development.

The United Nations Conference on Trade and Development (UNCTAD) classifies 29 countries as SIDS (see Table 8.1). There are ten in the Caribbean

Table 8.1 UNCTAD list of small island developing states

Antigua and Barbuda	Maldives	Samoa
Bahamas	Marshall Islands	Sao Tome and Principe
Barbados	Mauritius	Seychelles
Cape Verde	Micronesia (Federated States of)	Solomon Islands
Comoros	Nauru	Timor-Leste
Dominica	Palau	Tonga
Fiji	Papua New Guinea	Trinidad and Tobago
Grenada	Saint Kitts and Nevis	Tuvalu
Jamaica	Saint Lucia	Vanuatu
Kiribati	Saint Vincent & the Grenadines	

and the Americas, 12 in the Pacific, 5 in Africa and 2 in Asia. This represents a significant geographical dispersion and highlights the difficulties they face in forging common solutions to many of their inherent challenges. For example, while a similar template for marine management systems could be used by different small states, remoteness and geographic dispersion prevents the pooling of resources for their implementation. Therefore, the sustainable harvesting of marine resources presents a significant opportunity for environmentally sound, socially inclusive economic growth and development. Several countries and regions are starting to assess the need for and are striving to implement common governance frameworks under an ocean space approach (marine spatial planning). Examples of their use can be found in the design of a draft national marine policy in the Bahamas, the oceans economy road map of Mauritius and the regional oceans policy adopted by the Organisation of Eastern Caribbean States (OECS).

In the case of the Bahamas, an integrated marine policy framework is being implemented to manage the ocean space and marine resources with the objective to achieve greater clarity to the roles, functions and actions of approximately 34 governmental bodies. It will also hold competence over marine affairs and consolidate 36 pieces of subsidiary legislation regulating marine activities. Meanwhile, Mauritius launched its first oceans economy roadmap in 2013, which seeks to take advantage of the immense economic potential of oceans. The roadmap places emphasis on the need to make use of the untapped value locked up in the EEZ by ensuring sustainable and coordinated utilisation of living and non-living resources. Sectors of interest include tourism, seaports and seafood-related activities. In order to advance the roadmap's priorities, a national public–private taskforce as well as an oceans business park and an oceans research centre will be created. This new institutional set up will be complemented and supported by a comprehensive regulatory review of the oceans economy. Likewise, the member states of the OECS in 2013 approved the Eastern Caribbean Regional Oceans Policy. This regional policy architecture provides a framework that guides planning and development of marine activities in a rational and sustainable manner within the region.

It is too early to point toward concrete results of these policy frameworks, but a shift in mind-set towards a 'common oceans space' at a regional level is starting to emerge. Other SIDS could also consider the development of broader regional ocean governance architecture, based on the framework of the United Nations Convention on the Law of the Sea (UNCLOS), that ensures the sustainable management of living and non-living resources as an effective avenue to bolster economic growth and development. Such an approach could support common regulations and institutions aimed at governing enterprise and infrastructure development and investment. This approach would allow for the necessary consolidation of resources that might otherwise be out of reach for SIDS acting individually.

Addressing the challenges and seizing the opportunities presented by sustainable tourism in SIDS requires strong governance, including cross-cutting tourism policies, participatory frameworks and enhanced public–private partnerships. Further opportunities exist for SIDS to promote socio-economic growth, resilience and sustainability through regional integration and cooperation.

Governance, private-sector engagement and community development

SIDS need to establish specific governance and management structures for tourism, which engage with environment, culture, transport, trade and industry, and planning and development responsibilities, to enable a participatory approach and to foster meaningful partnerships between the public and private sectors and local communities.

The Coral Triangle Initiative

The Coral Triangle Initiative (CTI) illustrates how SIDS can strengthen the co-ordination of tourism policies and actions within groups of SIDS, to gain economies of scale in tourism management, to share information and knowledge, and, as integrated regional destinations, to achieve a higher profile in the marketplace (i.e. a brand identity).

The Coral Triangle (CT) area (see Figure 8.1) contains the most marine species on earth and supports the livelihoods of nearly 400 million people. Spanning 6 million square kilometres (2.3 million square miles) of the archipelagic waters of Indonesia, Malaysia, Papua New Guinea, the Philippines, the Solomon Islands and Timor-Leste, the CT has the highest coral and reef fish diversity globally and provides vital spawning grounds for whales, turtles and other charismatic species. In addition, one third of its people depend on these waters for their livelihoods. The CT has been called the 'Amazon of the Seas', as it is an epicentre of marine life abundance and diversity. As a global centre of marine biodiversity, the CT is a major conservation priority. Unsustainable fishing, rapid population growth, urbanization and the effects of climate change have threatened these abundant and valuable marine and coastal resources, and so an agreement was forged between the countries of the CT to work together to combat these threats and sustain this extraordinary environment and all it provides to human wellbeing.

Officially titled the Coral Triangle Initiative on Coral Reefs, Fisheries and Food Security, the CTI is a multilateral partnership of six countries formed in 2007 to address the urgent threats facing the coastal and marine resources of this biologically and ecologically rich region. The CTI is managed through a Secretariat based in Jakarta, Indonesia (World Ocean Council 2016). The focus is on empowering people to sustainably manage marine ecosystems so they can support their well-being, now and into the future. By connecting the efforts in-country to the wider region, they are ensuring that successes are amplified, creating momentum from the ground–up for changes that last.

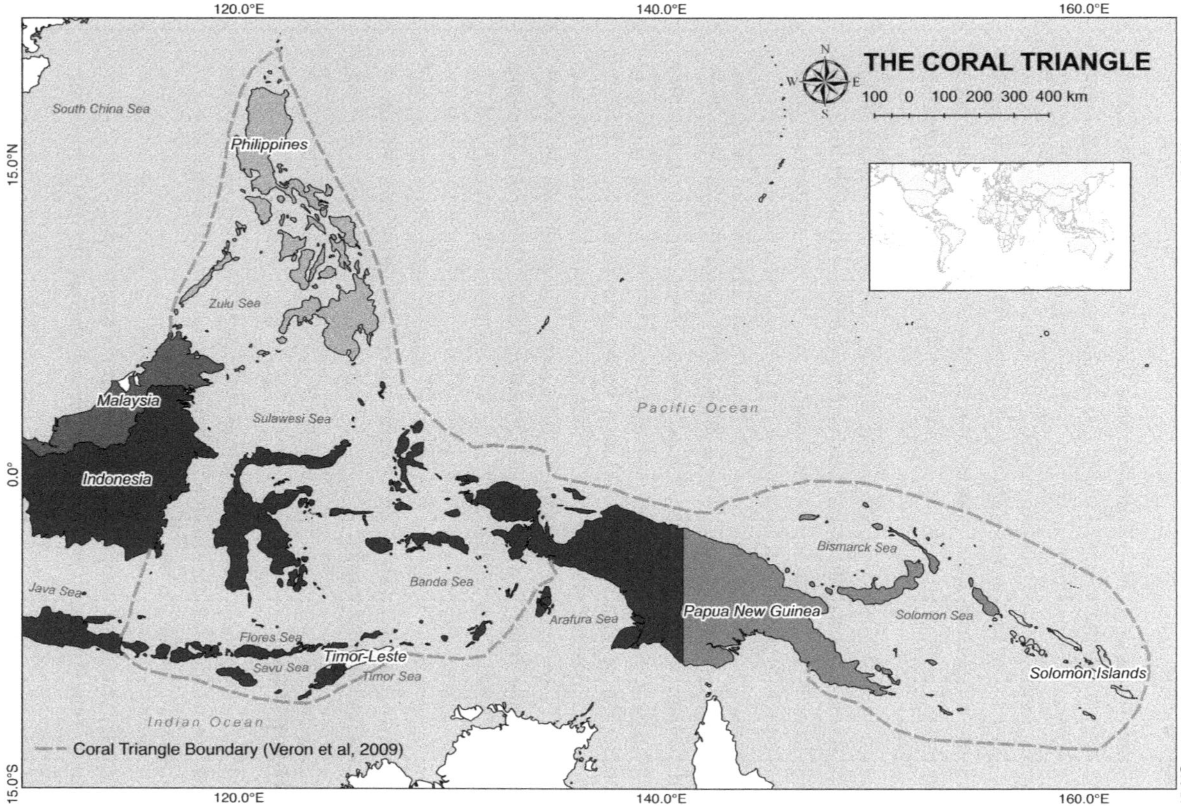

Figure 8.1 Map of the Coral Triangle

(*Source:* Coral Triangle Center, with permission.)

CASE STUDY: WESTERN PACIFIC OCEAN, THE CORAL TRIANGLE

Background

The Coral Triangle (CT) is a marine area located in the western Pacific Ocean. It includes the waters of Indonesia, Malaysia, the Philippines, Papua New Guinea, Timor-Leste and the Solomon Islands. Named for its staggering number of corals (nearly 600 different species of reef-building corals alone), the region nurtures six of the world's seven marine turtle species and more than 2,000 species of reef fish. Over 120 million people live in the CT and rely on its coral reefs for food, income and protection from storms.

Recognizing the need to safeguard the region's marine and coastal biological resources, a multilateral partnership, the Coral Triangle Initiative on Coral Reefs, Fisheries and Food Security (CTI), was formed in 2007. There is currently an interim Secretariat for the CTI, which is in the process of transitioning to a permanent Regional Secretariat. The interim Secretariat is based in Jakarta, Indonesia. Through the CTI, the CT countries have agreed to support people-centred biodiversity conservation, sustainable development, poverty reduction and equitable benefit sharing. The CTI Regional Plan of Action (RPOA) is a living and non-legally binding document, to conserve and sustainably manage coastal and marine resources within the CT region that takes into consideration laws and policies of each country. At the third Senior Officials Meeting (SOM3) in Port Moresby, Papua New Guinea in March 2009, the CT governments recognized the possible need to revisit Sections IV-VI in the Plan (Coordination Mechanisms and Implementation Partners, Financial Resources, and Monitoring and Evaluation), and tasked the CTI Coordination Committee to continue to advance work on these topics.

Action plans

The RPOA sets goals, targets and actions at the regional level, which either require or would benefit from collaborative action amongst the CT countries. The five goals are:

- Priority seascapes designated and effectively managed.
- Ecosystem Approach to Management of Fisheries (EAFM) and other marine resources fully applied.
- Marine Protected Areas (MPAs) established and effectively managed.
- Climate change adaptation measures achieved.
- Threatened species status improvement.

At the national level, each CT country has developed a National Plan of Action (NPOA), which specifies domestic priorities and actions for the CTI.

The Regional State of the Coral Triangle (RSCT) report provides benchmarks on the biophysical, governance and socio-economic attributes of the six CT countries, also known as CT6 (Indonesia, Malaysia, Papua New Guinea, the Philippines, the Solomon Islands and Timor-Leste), and the threats, vulnerabilities and emerging issues in each country and in the CT6 as a collective unit. Using the Driver–Pressure–State–Impact–Response (DPSIR) framework, the RSCT report explains the linkages among the higher level outcomes; and the goals, targets and actions which constitute the RPOA and the NPOA of the CTI. Specifically, the report summarizes the status of governance, and the biophysical and socio-economic attributes of CT6, and it highlights the unique social and ecological features of the CT, which show the great value of coral reefs and associated habitats. Furthermore, the RSCT report:

- Establishes a framework for tracking progress of the CTI toward goals set out in the RPOA.
- Identifies the information gaps and establishes the status of the coral reefs and associated habitats, fisheries and food security.
- Discusses the relationships of the ecological and social conditions of the CT6 countries; the threats they face; and their corresponding responses to the national, seascape and regional challenges and opportunities.
- Initiates the linkages between the NPOA and the RPOA with respect to the desired sustainable development outcomes of: (1) conserving the CT coral reefs and associated ecosystem functions, goods and services; (2) establishing sustainable fisheries; and (3) improving food security.

This regional report is a 'living document' that will have to be updated over time as the countries jointly address the various issues confronting them individually and as a group. This first edition of the RSCT report provides baseline data and information for monitoring the countries' progress in achieving the goals and higher level outcomes of the CTI. This RSCT report also highlights the CTI's inter-relational framework, specifically the state and pressures and the broader context of drivers. This report further discusses how the RPOA is linked to the desired higher level outcomes of coral reefs, fisheries and food security (CFF). While the RPOA envisions the attainment of five goals by 2020, the longer term vision for the CTI includes the following:

- To stabilize and/or maintain coral reef ecosystem integrity and services.
- To boost and sustain fish stocks.

- To improve the affordability, availability, quality and safety of food coming from coastal and marine environments.

Environmental status

The status of CFF was inferred from data and information presented in the Six-Country SCT reports; and supplemented with information from literature, both published and unpublished. Overall, coral cover in the CT6 countries had been on a declining trend since the 1980s to the mid-2000s. In the entire Indo-Pacific region, coral cover was recorded from 42.5% during the early 1980s to 22.1% by 2003. In the Philippines, coral cover had declined since the 1980s, while East Indonesia and Papua New Guinea had stable coral cover from 1984 to 2004. The coral reefs in Indonesia, Malaysia and the Solomon Islands were also in relatively better condition than in the Philippines. Papua New Guinea and Timor-Leste lacked data to evaluate the status of their coral reefs at the national level.

Reef fish biomass values are often not estimated in many parts of the CT6 countries and vary greatly across the CT6. In the Philippines, more than 50% of sites surveyed from 1991 to 2004 showed reef fish biomass of less than 10 tons per square kilometer (t/km^2). Unrestricted areas in Papua New Guinea had an average reef fish biomass of 12.7 t/km^2 and several reef areas in the Solomon Islands reported fish biomass greater than 100 t/km^2. The extent and status of mangroves and seagrasses at the national level are not fully known for most of the CT6 countries. Research related to these ecosystems has not kept pace with the knowledge and information required to ensure the sustainability of the coral reefs. Despite the declining health of the reefs, their value remains high. Overall annual net benefits, including tourism and other ecosystem services, are estimated to be no less than US$12 billion.

Over 120 million people are dependent on ecosystem functions, goods and services; and their combined contribution to the gross domestic product amounts to US$1.2 trillion, with capture fisheries valued at US$9.9 billion, representing 10.5% of the global market. Major drivers common across the CFF were identified during a regional workshop attended by government officials from each of the CT6 countries: (1) population growth, (2) cultural challenges, (3) education, (4) coastal development, (5) poverty and governance, (6) demand for fish, and (7) climate change.

Other major stressors are excessive nutrient inputs and pollution, land and coastal development, and exploitation of threatened species. Consistent with the regional and country SCT reports, Reefs at Risk in the Coral Triangle (WWF 2008) identified overfishing and destructive fishing as the primary threats to coral reefs in the region, followed by watershed-based pollution, coastal development, and marine-based pollution and damage.

Tracking goal progress

Much of the efforts in the CT6 countries appear to be focused on establishing and managing MPAs, as well as applying the EAFM and other marine resources. In Indonesia, Papua New Guinea and the Philippines, at least two-thirds of their efforts, as articulated in their NPOAs, deal with these two goals. The NPOAs of Malaysia and Timor-Leste also provide for considerable investments in EAFM. Timor-Leste considers MPAs and climate change adaption measures among its top priorities. A CTI Index was developed and piloted during the RSCT report and CTI Monitoring and Evaluation Working Group (MEWG) meetings in Jakarta in October 2012.

The CTI Index indicates the extent of accomplishment of the five CTI goals without making connection to the higher level outcomes. The scores for CFF activities in the CT6 countries and in the region averaged 42%. Nine priority actions agreed upon by the CT6 countries best gauge their progress. Based on the presentations at the Senior Officials Meeting (SOM) in October 2011, the actions were categorized as (1) completed, (2) in progress, and (3) not started. An action was deemed completed at the level of the Technical Working Group (TWG) but may or may not indicate a formal endorsement at the SOM. All but two of the nine actions had either been started or completed, with the Region-Wide Early Action Plan for Climate Change Adaptation and the Coral Triangle Marine Protected Area System (CTMPAS) framework already completed. Two actions that pertain to threatened species status improvement had not been initiated.

Many gaps remain in knowledge and information crucial to the strengthening and capacity building of the CTI processes and systems, and standards are not yet in place for the higher level outcome indicators. However, opportunities exist for the missing information to be collected by consolidating the monitoring programs of different groups or organizations, and developing protocols to synthesize available fragments of information to arrive at national estimates.

The RSCT report allows better transparency for countries to track the progress of their activities (e.g. through the CTI Index). Ways to link these activities had been proposed in the CT6 NPOAs and CTI RPOA, and desired outcomes had been elicited from the DPSIR approach. It is, therefore, suggested that the RSCT be updated on a 3-year basis, as an integral task of the CTI MEWG. In the next phase of the CTI, five major thematic thrusts and 13 action areas were proposed for consideration by the CT6, the Regional Secretariat and the development partners.

Major thematic thrusts and action areas proposed by the CTI

- Achieve synergies at different governance scales to earn the value-added benefits of overcoming transactional costs (e.g. improving seascapes and operational functions of the CTI as a result of cooperation and complementation):
  - — coordinate actions through improved processes, systems and standards, such as awards and incentive systems for best practices across MPAs and MPA networks and social enterprises;
  - — ensure that benefits from institutional coordination are plowed back into managing ecosystems and their uses through sharing agreements; and
  - — monitor the costs and benefits of cooperative governance to gauge impacts on human and ecological systems; and provide timely response feedback systems, including enabling conditions for social enterprise development.
- Invest in capacity building and knowledge management to overcome the lack of governance capabilities in CTI systems, processes and standards (e.g. CTMPAS and EAFM):
  - — build the resiliency and capacity of local, national and regional bodies in planning and implementing the CTI NPOA/RPOA (e.g. incentives through conditional grants linked to incentives-based progress of capabilities and performance);
  - — understand and apply science-based learning through adaptive research and development learning networks (e.g. Coral Triangle Center, Coral Triangle Initiative – Coastal Learning Adaptation Network); and
  - — organize monitoring and feedback-sharing forums for the regular updating of the country and regional SCT reports at least every 3 years.
- Exchange resources and engage and empower equitable partnerships:
  - — establish a CT regional investment fund that will rationalize financial and economic support for the CTI; and develop mechanisms that will ensure the sustainability of the CTI, including public–private partnerships; and
  - — improve access of vulnerable coastal communities to available food resources and social enterprise development.
- Commit to the harmonization of fisheries production targets with biodiversity conservation and food security needs:

- — complete red list and critical habitat assessments; and harmonize these at local networks and seascapes, integrating EAFM and CTMPAS (e.g. Sulu–Sulawesi Marine Eco-region (SSME) and Bismarck–Solomon Seas Marine Eco-region);
- — establish safety nets and diversify livelihoods that promote fisher stewardship, such as through conditional cash transfer programs; and
- — ensure that international and local agreements consider traditional ecological knowledge and wisdom and customary marine tenure through knowledge management and sharing forums linked to regional organizations.

- Reduce risks and threats through integration of the Local Early Adaptation Plan and the Region-Wide Early Action Plan:
  - — form regional climate adaptation and disaster risk reduction response programs (e.g. CTI climate research and development sharing exchanges with other regional forums); and
  - — mitigate and minimize threat–transfer effects, such as from illegal, unreported and unregulated fishing; and transmigration through joint enforcement agreements, such as the SSME learning shared on a CT region-wide scale.

These action areas were considered by the CT6 individually as well as collectively by the various technical working groups that are involved in the preparation of the detailed RPOA and NPOA. The NPOA and technical working groups will assign responsibilities and budgets. The establishment of the CTI Regional Secretariat is estimated to cost US$3.5 million, and efforts toward attaining the five CTI goals would require US$4.9 million. Although these figures seem huge, the total of US$8.4 million is less than 1% of the capture fisheries value of the CT6 countries, which was estimated at US$9.9 billion in 2007. Continuing to invest in the CTI is a worthwhile endeavour; and regional cooperation and coordination among the CT6 countries are essential for attaining CTI goals and desired higher level outcomes.

The Australian government support program for the CTI

The Australian government has committed to a phased multiyear program of support to the CTI that will evolve and respond as the initiative grows. The first stage of support focused on building foundations and momentum for the CTI. The current second stage of support is focused on building the capacity of individuals and communities to develop and grow sustainable industries and livelihoods, by drawing on both local and regional natural assets. A third stage of support was announced in November 2014 and is currently being programmed in consultation with the CTI member countries. Areas of focus for the current phase include:

- Helping to build the CTI into an effective regional forum.
- Helping to achieve transformational changes to the management and use of marine resources.
- Developing sustainable livelihoods.
- Improving knowledge for better planning and decision making.
- Supporting communities to be effective stewards of their environment and natural resources.
- Building skills and capabilities for better marine and coastal management.

Principles for Australian government CTI support

In planning and providing support, the Australian government has, and will, apply the following principles:

- Support will target people-centred biodiversity conservation, sustainable development, poverty reduction and equitable benefit sharing.
- Support will match the pace of the CT6 and allow sufficient time and opportunity to ensure meaningful engagement.
- Support will match standards for engagement in fragile states and recognize capacity constraints.
- Support will recognize the stage of the development of the CTI as a regional forum. Priority will be given to activities that will underpin the long-term strength and success of the CTI.
- Support will respect and encourage the role of national governments, and be based on detailed, country-led planning, national government priorities, and consultation with the national governments.
- Support will respect and support the role of the CTI Regional Secretariat, Senior Officials meetings, and Council of Ministers in directing and coordinating achievement of CTI goals.
- Support will build on, leverage and coordinate with other complementary programs and activities of the Australian government and initiatives in the region, and will capitalize on Australia's areas of expertise and engagement.
- Support will be aligned with relevant Australian aid program Country Strategies and Partnerships for Development, or regional strategies and commitments (such as the Pacific Plan and Cairns Compact). Overseas

Development Assistance-funded activities under this Support Plan will be subject to Australian aid program processes and requirements directed at ensuring effectiveness.
- Support will be communicated actively and openly in the interests of improved donor contribution.

Conclusion

There is broad scientific consensus that the CT represents a global epicentre of marine biodiversity. Spanning just 1.6% of the planet's oceans, within the CT region are 76% of all known coral species, 37% of all known coral reef fish species, 53% of the world's coral reefs, the greatest extent of mangrove forests in the world, and vital spawning and juvenile growth areas for tuna and other globally-significant commercial fish species. The countries of the CT region recognize that their marine and coastal resources are a cornerstone of their economies and societies, and that these resources are under increasing threat. Increasing pressures include overfishing, unsustainable fishing practices, land-based sources of marine pollution, coastal habitat conversion and climate change.

A range of Australian stakeholders have an interest, or are involved in, activities which support and compliment the CTI. These include research and educational institutions, universities, non-government organizations, community organizations, state/territory governments and local governments. The experience, knowledge and expertise of these stakeholders are valuable, and the Australian government will seek to engage with these stakeholders in planning and implementing support for the CTI. When the CTI was established, a group of stakeholders, including the Australian Government, were invited to become Partners to the CTI.

Partners have been asked to provide funding support, as well as technical and strategic expertise and advice, to assist the pursuit of CTI goals and to facilitate the establishment of the CTI as a fully functioning, effective and representative regional forum. The other seven Partners are: Asian Development Bank, Conservation International, Coral Triangle Center, Global Environment Facility (GEF), The Nature Conservancy (TNC), United States Government (USAID), and World Wide Fund for Nature (WWF).

The CTI case study illustrates a collaborative agreement and effective partnership between government and stakeholders. Global challenges demand global solutions. Most of the challenges facing sustainable development in SIDS are cross-cutting in nature and require an integrated approach and new governance models at both national and international levels, engaging public and private sectors as well as civil society.

Discussion

The UN Steering Committee on Tourism for Development (SCTD) brings together the experience and expertise of each member to create synergies and a more coordinated, effective and efficient delivery of technical assistance in tourism to developing countries, complementing their efforts to build up a competitive national tourism sector. The Global Partnership for Sustainable Tourism is a global initiative launched in 2011 to promote sustainability in the mainstream of tourism policies, development and operations. It emerged as a more permanent successor to the International Task Force on Sustainable Tourism Development (ITF-STD). Sustainable tourism is among the five initial programs to be developed within the 10-Year Foresight Process, opening important opportunities for the UN system and member states to advance SIDS-specific lines of action that support sustainable tourism in those countries. The United Nations World Tourism Organization (UNWTO) works in close cooperation with the World Travel and Tourism Council (WTTC) to promote tourism as a main driver of economic development and sustainability. WTTC gathers the Chairs, Presidents and CEOs of 100 of the world's foremost Travel and Tourism companies.

References and further reading

Asian Development Bank (2014). *Regional State of the Coral Triangle: Coral Triangle Marine Resources: Their status, economies, and management*. Philippines: Asian Development Bank.

CTI-CFF Regional Secretariat (2016). *Regional Plan of Action: Coral Triangle Initiative on Coral Reefs, Fisheries and Food Security (CTI-CFF)*. Indonesia: CTI-CFF Regional Secretariat.

UNCTAD (United Nations Conference on Trade and Development) (2014). 'The oceans economy: Opportunities and challenges for small island developing states'. http://unctad.org/en/PublicationsLibrary/ditcted2014d5_en.pdf (accessed: April 2017).

UNDP (United Nations Development Programme) and UN-OHRLLS (United Nations Office of the High Representative for the Least Developed Countries, Landlocked Developing Countries and Small Island Developing States) (2015). 'Financing for development and small island developing states: A snapshot and ways forward'. Discussion Paper. www.undp.org/content/undp/en/home/librarypage/poverty-reduction/FfD-SIDS-UNDP-OHRLLS.html (accessed: April 2017).

UNEP (United Nations Environment Programme) (2014). 'Emerging issues for small island developing states: Results of the UNEP/UN DESA Foresight Process'. https://wedocs.unep.org/rest/bitstreams/16063/retrieve (accessed: April 2017).

UNEP (United Nations Environment Programme) (2014). 'Guidance manual on valuation and accounting of ecosystems services for small island developing states'. Regional Seas Reports and Studies No.193. www.openchannels.org/sites/default/files/literature/Guidance%20Manual%20on%20Valuation%20and%20Accounting%20of%20Ecosystem%20Services%20for%20Small%20Island%20Developing%20States.pdf (accessed: April 2017).

UN-OHRLLS (United Nations Office of the High Representative for the Least Developed Countries, Landlocked Developing Countries and Small Island Developing States) (2014). 'Small island developing states: Small island big(ger) stakes'. http://unohrlls.org/custom-content/uploads/2013/08/SIDS-Small-Islands-Bigger-Stakes.pdf (accessed: April 2017).

UNWTO (United Nations World Tourism Organization) (2014). 'Tourism in small island developing states (SIDS) – Building a more sustainable future for the people of islands'. www.e-unwto.org/doi/pdf/10.18111/9789284416257 (accessed: April 2017).

World Ocean Council (2016). 'Marine spatial planning: Case studies'. http://oceancouncil.org/wp-content/uploads/2016/05/WOC-MSP-Case-Studies-Mar-2016.pdf (accessed: April 2017).

WWF (World Wide Fund for Nature) (2008). 'Reefs at risk in the Coral Triangle'. http://wwf.panda.org/?147522/Reefs-at-Risk-in-the-Coral-Triangle (accessed: April 2017).

Chapter 9

Ocean governance and the blue economy

The last frontier

The New Atlas of Planet Management was regarded as the most groundbreaking survey of the state of our planet when it was first published in 1984. After over 20 years in print it was revised in 2005 and in this publication Carl Safina made reference to the oceans as the 'last frontier' on earth (Safina 2005, p.74). It is widely accepted that oceans shall decisively influence world geopolitics, globalization, socio-economic balance, and life on earth. Today, 82% of all nations border the world's oceans or seas. Of these, some 40 nations (30%) are islands. The world's oceans have afforded the human populace freedom to explore and acquire natural resources and wealth distributed over 29% of the land. With 71% of the earth's surface covered by water, the oceans encompass not only two-thirds of the world surface but could well impact the lives of three-quarters of the world's population by 2030.

The new legal regime for the oceans has established zones of maritime jurisdiction and specified rights and responsibilities for all states including landlocked states. The maritime zones that define the oceanic realm cover a geographic area where both the land and the water interact through the dynamics of natural processes. The boundary of these maritime zones encloses a space, which extends from the ocean across the land–water interface to an inland limit. It is difficult defining such an all-embracing area in terms of physical or administrative boundaries.

Ten principles for high seas governance

The ocean is a unique, extraordinary and vital element of our planet. It sustains life on Earth by generating oxygen, absorbing carbon dioxide from the atmosphere, regulating climate and temperature, and providing a substantial portion of the global population with food and livelihood. It provides medicine, energy and transport routes, amongst many other services, and has been a nexus for various cultures. National jurisdiction over resources has recently been extended to 200 nautical miles in the sea, leaving about 60% of the ocean as 'high seas' and deep seabed beyond national jurisdiction. Long treated as though empty of life other than some fishes, we are now discovering that these areas of the oceans contain some of the richest biodiversity on the planet. This biodiversity is under increasing threat from many sources, including irresponsible fishing and shipping activities, pollution and climate change. Yet there is no comprehensive policy or management framework to govern the high seas, and the patchy laws that exist are largely based on 17th century principles of open access, ignoring many of the environmental principles that have been long applied for land and atmosphere, and even for outer space.

Beyond national jurisdiction, many human activities remain unregulated; ecosystem considerations are seldom taken into account and scientific information is often ignored. Experience in managing ocean resources has provided many lessons that need to be put into practice. Critical policy issues to improve high seas governance are ready to be addressed and practical steps can be taken to improve biodiversity conservation. Past limitations in monitoring and enforcement of ocean governance regimes have been overcome by technological advances, but the political will to commit resources to this task is still lacking. A good start has been made, but more needs to be done. We stand at a critical time. The ocean governance system must evolve and the following modern principles applied to improve high seas management and ensure sustainable development of the world's oceans.

Conditional freedom of activity on the high seas

The time of treating the oceans as 'open access, common pool' resources are over. Our ocean resources have proven to be exhaustible, so their use needs to be regulated. There is a need to reaffirm and enforce international law, in particular United Nations Convention on the Law of Sea (UNCLOS), and to condition the enjoyment of high sea's freedoms upon the implementation of the convention's duties. Access to common ocean resources must be twinned with comprehensive and effective governance that includes monitoring, sanctions and enforcement.

Protection and preservation of the marine environment

The obligation to protect and preserve the marine environment, including in areas beyond national jurisdiction, is enshrined in many legal instruments at the global and regional levels, but this obligation has largely not been met. Today, an increasing number of threats make it urgent to ensure such protection. There is a pressing need to deliver the World Summit on Sustainable Development targets and commitments towards ensuring the sustainable development of the world ocean, including those parts that lie beyond national jurisdiction.

International cooperation

Biodiversity beyond national jurisdiction cannot be protected effectively unless countries act together. For conservation and management measures to be effectively implemented, coordination is essential across sectors and states. Institutional mechanisms for ensuring such cooperation are required. The current model that is based on regional fisheries management organizations and agreements needs to be refined to truly manage the oceans and not simply to allocate scarce and overused resources. Clearly it is time for states to cooperate to expand the mandate of existing regional management organizations and agreements and to fill gaps to ensure the conservation and sustainable and equitable use of all high seas marine resources.

Science-based approach to management

Management decisions should be based on the best available science and not on the lobbying of a powerful few. Further scientific research is required to underpin ecologically sustainable resource use, to inform the adaptive application of conservation measures, and to enhance the development of criteria for monitoring. Scientific knowledge, understanding and awareness of the high seas and deep seabed need to be improved rapidly by building on and going beyond the various projects and initiatives currently undertaken. As conservation measures are applied in areas of resource use that have a commercial value, decisions relating to areas beyond national jurisdiction should also consider socio-economic information to encourage effective compliance and application of economic incentives.

Public availability of information

We are just starting to learn more about the biodiversity of the high seas and deep seabed. To adequately manage these resources it is critical to increase knowledge of our ocean and the impacts of resource use and to make it available to the public. Information on the use of marine resources, in particular data from users, scientific observations and research results should be exchanged and made freely available. Such information should not be kept in the jargon that is only understood by a few, but should be made available in a language that is accessible to all. Widespread popular support and increasing public awareness about ocean resources and those who benefit from them are critical to ensuring that conservation and management efforts are successful and sustainable.

Transparent and open decision-making processes

Societies are demanding more effective management of fisheries and marine ecosystems and there is an urgent need to ensure greater transparency and increased participation by stakeholders in managing high seas resources. In addition, it is critical that decision-making processes are conducted in a manner that is transparent and accountable to minimize the likelihood of disputes and to promote international cooperation. Regional and global organizations need to promote the meaningful participation of all interested stakeholders in decision making, provide observers access to all meetings and documents, and receive advice from all interested observers.

Precautionary approach

The knowledge base for managing the high seas is arguably weaker than for other ecosystems, but this should not be used as ground for lack of action. Adaptive management allows decision making to respond to changes and inherent levels of uncertainty. In high seas governance, more than any other place, the precautionary principle should become standard practice. Environmental assessment requirements and the precautionary approach should be operationalised for all high seas activities and should be used as a globally applicable default mechanism for existing, emerging and new activities on the high seas. This will require placing the burden of proof on those who argue that an activity will not cause significant harm to show that this is so, and make the responsible parties liable for environmental harm.

Ecosystem approach

As was recognized by the UN General Assembly more than 35 years ago, 'the problems of ocean space are closely inter-related and need to be considered as a whole'. There is a strong need to consider the bigger picture, to give meaning to ecosystem management and to move away from the sectoral and species-based approaches from which ocean governance systems have long suffered. Ecosystem approaches need to be further refined and made operational. Large-scale marine spatial planning and networks of marine protected areas, and other area-based management measures for biodiversity conservation purposes, should be integral parts of an ecosystem approach to fisheries and oceans management. The high seas offer the ideal place to implement ecosystem management with fewer political boundaries, fewer stakeholders and virtually no property rights.

Sustainable and equitable use of ocean resources

Using ocean resources in areas beyond national jurisdiction requires a balance between the rights and interests of individual users and those of the international community. Management of resources should result in such resources being used in a sustainable manner to maintain the biological diversity to meet the needs of present and future generations. There is a need to discriminate against harmful practices and perverse incentives, especially destructive practices that adversely impact biodiversity beyond national jurisdiction. There is also a strong need to promote and foster international cooperation and the development, adoption, and transfer of environmentally sound technologies on an equitable and mutually acceptable basis. Particular attention should be given to benefits to and the interests of developing countries.

Responsibility of states as stewards of the global marine environment

States need to bear responsibility for activities in the high seas by their own governmental agencies, by vessels under their flags, and by their nationals, both individual and corporate. States are responsible for assuring that national activities are carried out in conformity with international law and with the above-mentioned principles. The activities of ships and nationals in the high seas should require authorization and continuing supervision and monitoring by the appropriate state. In accordance with the polluter/user pays principle, states should be liable to other states and to the global community in case of damage to the marine environment and resources caused by their vessels and nationals.

Marine biodiversity and ocean temperature

As was mentioned briefly in Chapter 2, a team of scientists mapped and analyzed global biodiversity patterns for over 11,000 marine species ranging from tiny zooplankton to sharks and whales (Tittensor et al. 2010). The researchers found striking similarities among the distribution patterns, with temperature strongly linked to biodiversity for all 13 groups studied. These results imply that future changes in ocean temperature, such as those due to climate change, may greatly affect the distribution of life in the sea. The scientists also found a high overlap between areas of high human impact and hotspots of marine diversity. Much research has been conducted on diversity patterns on land, but our knowledge of the distribution of marine life has been more limited. This has changed through the decade-long efforts of the Census of Marine Life.

In the Census of Marine Life, the researchers synthesized global diversity patterns for major species groups including corals, fishes, whales, seals, sharks, mangroves, seagrasses and zooplankton. In the process, the global diversity of all coastal fish species has been mapped for the first time. The researchers were interested in whether there are consistent 'biodiversity hotspots' (i.e. areas of especially high numbers of species for many different types of marine organisms simultaneously). They found that the distribution of marine life showed two fundamental patterns: coastal species such as corals and coastal fishes tended to peak in diversity around Southeast Asia, whereas open-ocean creatures such as tunas and whales showed much broader hotspots across the mid-latitude oceans. The scientists also tested whether these global patterns could be consistently explained by one or more environmental factors.

Temperature was the only factor found to be linked with the distribution of all species groups, with the availability of habitat also playing a role. The researchers observed that it was striking how consistently temperature was linked with marine diversity. This relationship suggests that ocean warming, such as that due to climate change, may rearrange the distribution of oceanic life. They noted that while we are increasingly aware of global gradients in diversity and their associated environmental factors, our knowledge of patterns in the ocean has lagged behind that of patterns on land. The Census of Marine Life attempts to help overcome this disparity. The study also assessed the overlap between hotspots of marine diversity and human impacts, i.e. the combined effects of fishing, habitat alteration, climate change and pollution. Human impacts were found to be particularly concentrated in areas of high diversity, suggesting the potential for severe species losses in these regions. They also noted that the combined effects of exploitation, habitat alteration, pollution and climate change are threatening the diversity of life in the global ocean.

The research provides further evidence that limiting ocean warming and other human impacts will be particularly important in securing these hotspots of marine biodiversity into the future. The Census of Marine Life also highlighted the need to maintain biodiversity in the face of these impacts. Biodiversity and the functioning of ecosystems are often tightly coupled, with highly diverse ecosystems providing more goods and services that benefit human beings, as well as being more resilient in the face of disturbance, than less diverse ecosystems. The observed concentration of human impacts in our richest marine areas is a worrying indication of our growing footprint in the oceans. Much of the data used for this study come from the Ocean Bio-geographic Information System (OBIS), a public database created by the Census of Marine Life. With OBIS, a framework has been created for sharing and reusing data, which makes this type of global, all-encompassing science possible.

Marine biodiversity and climate change

Research into the impact of climate change has found that warming oceans will cause profound changes in the global distribution of marine biodiversity. In a study published in the journal *Nature Climate Change* (Molinos et al. 2015), an international research team modeled the impacts of a changing climate on the distribution of almost 13,000 marine species, more than 12 times as many species as previously studied. The study found that a rapidly warming climate would cause many species to expand into new regions, which would impact on native species, while others with restricted ranges, particularly those around the tropics, are more likely to face extinction. A study by the ARC Centre of Excellence for Coral Reef Studies at the University of Queensland offered hope in the potential of species to track and follow changing climates, but it also gave cause for concern, particularly in the tropics, where strong biodiversity losses were predicted. This is highly germane to Australia's coral reefs, because complementary studies have shown high levels of extinction risk in tropical biotas, where localized human impacts as well as climate change have resulted in substantial degradation.

To model the projected impact of climate change on marine biodiversity, the researchers at the ARC Centre of Excellence for Coral Reef Studies used climate–velocity trajectories, a measurement which combines the rate and direction of movement of ocean temperature bands over time, together with information about thermal tolerance and habitat preference. According to the study, the analysis provides the simplest expectation for the future distribution of marine biodiversity, showing recurring spatial patterns of high rates of species invasions coupled with local extinctions. The researchers say this will make currently distinct ecological communities much more similar to each other in many regions by the end of the century. The researchers warned that the resultant novel combinations of resident and migrant species will present unprecedented challenges for conservation planning. Above all, this study showed the broad geographic connections of the effects of climate change.

Conservation efforts need to be facilitated by cooperation among countries to have any real chance of combating the potentially severe biodiversity losses that a changing climate might impose – and Africa is no exception. As an example for a region of its size, the coastal and marine environment around South Africa is regarded as one of the most varied in the world, largely due to the dynamic presence of two major currents. The Agulhas Current flows strongly to the southwest along the eastern and southern coasts, bringing warm water and diverse tropical species from the Indian Ocean. The Benguela Current flows northward along the west coast, with wind-driven upwelling close inshore permitting cold nutrient-rich water to rise to the surface, and giving rise to productive coastal ecosystems and fisheries.

2050 Africa's Integrated Maritime Strategy

Africa faces many challenges against a background of civil wars, corruption, poor governance, and weaknesses in implementing strategies and policies. This is exacerbated by a lack of funding and inadequate investment coupled with an absence of strong private sectors in the African maritime economies. Africa's inland waters, oceans and seas are under pressure. Over the years, traditional maritime activities, such as shipping or fisheries have intensified, while new ones, such as aquaculture or offshore renewable energy, have emerged. However, the rise in intensity of activities at sea is taking place against the backdrop of insecurity, various forms of illegal trafficking, degradation of the marine environment, falling biodiversity, and aggravated effects of climate change.

Africa's coastal zones and ocean spaces bear certain unique features, as different from those of the other zones of the Indian Ocean region. One such maritime characteristic is that the continent is virtually flanked by sea on all four sides. Such uniqueness of the region has both positive and negative consequences for the coastal countries of Africa. The adverse consequences of the maritime topography, such as piracy and illegal fishing, are rather more seriously felt by the regional states. The unstable and fragile political regimes of many of the African littoral countries compound the problems of managing their maritime domains. Maritime criminal and illegal operations are confined not only to the coastal states but also to the island states of the continent. The inability of these states to combat the threats regularly posed by maritime non-state actors has resulted in the enormous naval militarisation of the African waters by foreign naval forces, Western and non-Western.

How far and how long the states of the region should depend on foreign countries for ensuring the safety of their coastal zones will also determine the level of independence that these states retain to keep their maritime wealth and domain under their sovereign control. In the past decades direct aggregate losses of revenue from illegal activities in Africa's Maritime Domain (AMD) amount to hundreds of billions US dollars, not to mention the loss of lives. The 2050 Africa's Integrated Maritime Strategy (2050 AIM Strategy) provides a broad framework for the protection and sustainable exploitation of the AMD in order to create wealth.

Definitions included in the 2050 AIM Strategy

- *Africa's Maritime Domain:* AMD refers to all areas and resources of, on, under, relating to, adjacent to, or bordering on, an African sea, ocean, lakes, intra-coastal and inland navigable waterways, including all African maritime-related activities, infrastructure, cargo, vessels and other means of conveyance. It also includes the air above the African seas, oceans, lakes, intra-coastal and inland navigable waterways, and also to the oceans' electromagnetic spectrum.
- *Combined Exclusive Maritime Zone of Africa (CEMZA):* Without prejudice to maritime zones as established by the UNCLOS for individual nations, CEMZA defines a common maritime zone of all African Union (AU) member states. It is to be a stable, secure and clean maritime zone in the view of developing and implementing common African maritime affairs policies for the management of African oceans, seas and inland waterways resources, as well as for its multifaceted strategic benefits. CEMZA will grant Africa enormous cross-cutting geostrategic, economic, political, social and security benefits, as well as minimize the risks of all transnational threats including organized crime and terrorism in Africa.
- *Inland waterways:* Refers to any navigable rivers, creeks, lakes, tidelands, lagoons, below-water baseline, or channel leading into such place having facilities for ships to moor and load or discharge, including offshore cargo handling facilities, harbours, berths, jetties, pontoons or buoys and wharves within the limits of the inland waterways in any place in a country, and includes any place declared to be an inland waterway under relevant national legislation.
- *Maritime safety:* Focuses on enhanced sustainable socio-economic development, the condition that reflects the ability of public and private entities to conduct legitimate activities such as territorial protection, resource extraction, trade, transport and tourism, free of threats or losses from accidents, negligence, natural and manmade disasters, for an integrated and prosperous Africa.
- *Maritime security:* Concentrates on improving sustainable socio-economic development, the condition that reflects the freedom of public and private

entities to conduct legitimate activities such as the exercise of sovereign and jurisdictional rights, resource extraction, trade, transport and tourism, free of threats or losses from illegal acts or aggression, for an integrated and prosperous Africa.

- *Maritime prosperity:* The condition that describes the actual value added by maritime activities measured in resources harvested, wealth created, investment, and dignity of African citizens. Maritime prosperity will entail cooperation and coordination at all regional levels to ensure that it is enjoyed by the African people.
- *Maritime sustainability:* The condition that describes the ability to conduct maritime activities over time in terms of infrastructural development and maintenance, equipment readiness, and equilibrium of renewable resources for an integrated and prosperous Africa.
- *Maritime viability:* The composite value of the four preceding interconnected maritime conditions (safety, security, prosperity and sustainability). This value will be applied by AMD stakeholders to complete assessments, conduct SWOT (strengths, weaknesses, opportunities and threats) and gaps analysis, and execute capacity-building programs to reduce all the threats, vulnerabilities and losses, and to achieve increased benefits and dignity, for an integrated and prosperous Africa.
- *Region:* Refers to the regional components of the AU as recognized by Resolution CM/Res.464 (XXVI) of the Council of Ministers of the Organization of African Unity on the division of Africa into five regions, namely: Northern, Western, Central, Eastern and Southern Africa.
- *Security:* A sea-centric approach to security encompasses all the traditional, state-focused, regional and continental notions for the protection of the states and regions of the continent. This protection can be achieved by all appropriate means. The emphasis is on collective security. Collective security embraces human security: national developmental aspirations and economic rights; and the right of popular participation in the decision-making process and other development issues. The challenge therefore will be how to develop and access resources and the basic necessities of life: the right to protection against threats such as famine and poverty; access to education and health facilities; protection against marginalization on the basis of gender; and vigilance against ecological and environmental degradation. Within the scope of this AIM Strategy, the focus will be on the key issues of maritime governance, maritime safety and security, maritime prosperity and maritime viability.

Decade of African Seas and Oceans

On 25 July 2015 Africa celebrated the African Day of Seas and Oceans under the theme, 'Harnessing the Blue Economy in Achieving the African Union Agenda 2063' and launched the 2015–2025 Decade of African Seas and Oceans.

The events took place as the 22nd Ordinary Session of Heads of States and Governments concluded in Addis Ababa, Ethiopia, at AU headquarters. The event raised awareness about challenges facing African seas and oceans and highlighted the 2050 AIM Strategy and Africa's blue economy. The Ordinary Session adopted the 2050 AIM Strategy and its plan of action, including a roadmap for the incremental implementation of the strategy in line with international maritime law.

The 2050 AIM Strategy includes a framework for action on, *inter alia*: fisheries and aquaculture; environmental and biodiversity monitoring; marine tourism; disaster risk management (DRM); handling and shipment of hazardous materials and dangerous goods; maritime governance; flag state and port state control; and illegal activities, including money laundering, piracy, maritime terrorism and human trafficking and smuggling by sea.

The author is of the opinion that an African Network of Maritime Clusters (ANMC), developed over a period of time, could facilitate collaborative economic development opportunities between the developed and developing maritime nations by linking the ANMC with the European Network of Maritime Clusters (ENMC). Nearly all the port cities dotted along Africa's extensive coastline are undergoing upgrading and expansion. There is huge potential for Africa's coastal nations and developing countries to collaborate with the developed countries, especially for trade and tourism. Now that there is peace in Angola and Mozambique, the full potential of their significant coastal and maritime resources are ripe for responsible development.

Special Purpose Vehicles financial circles

Special Purpose Vehicles (SPVs) and Special Purpose Entities (SPEs) have been household words within financial circles for decades. Lately, these entities have assumed centre stage in the wake of the worldwide financial crisis and its continuing after-effects. Although these vehicles have been under increased scrutiny, they can be highly beneficial when effectively managed. These vehicles play a vital role in the efficient operation of global financial markets.

Large corporations use these vehicles to meet specific objectives by way of obtaining finances, transferring risk and performing specific investment activities. In the past 30 years, SPVs have been used extensively as a means of securitisation for property-based financial products. This was further facilitated by easy credit conditions and financial deregulation. A period of growth in the world economy followed, driven by the use of these vehicles and related conditions. These added to the housing and credit bubble that would soon burst. The onset of the economic crisis in 2007 stressed the robustness of these vehicles and revealed a number of inherent defects in them. The flexibility of these vehicles meant that they were vulnerable to misuse with potentially devastating consequences. This has been proven time and time again by corporations such as Enron and Lehman Brothers, which were decimated due to inappropriate usage of these entities.

SPVs have a number of key utilitarian features and benefits that allow investors access to investment opportunities which would otherwise not exist. These include facilitating and supporting securitisation, financing, risk sharing and raising capital, to name a few. In the absence of SPVs, these objectives would not be possible without putting the entire corporation at risk. They also provide significant benefits to the parent firm by allowing ease of asset transfer, reducing 'red tape', providing tax benefits and legal protection.

Some of the key risks SPVs pose to the sponsoring firm are lack of transparency, reputational risk, liquidity and funding risk and equity risk. Lessons learnt from these collapses and their knock-on effects have led regulators to take strong measures that subject these vehicles to even more scrutiny than before. Stronger governance, increased oversight and more transparent reporting are some of the measures that form part of a tighter regulatory framework governing SPVs. A key example of this is the BASEL III regulation forcing banks to maintain adequate levels of capital and liquidity. Do the benefits of SPVs justify the risks involved? A broad range of views exist on this question with the middle ground being preferred by most. The future of SPVs depends on the ability to offer clarity to investors and to constantly balance risk.

Ocean economy SPVs

For the ocean economy, SPVs refer to an appropriate vehicle that will facilitate and encourage cooperation and collaboration within a defined multiuser ocean space (maritime domain). Maritime clusters are an appropriate SPV for onshore collaboration. Councils are an appropriate SPV for the ocean business community whereby ocean stakeholders collaborate in order to increase their comparative advantages via collective bargaining, lobbying and advocacy. Alliances are appropriate SPVs for responsible ocean governance where countries need to collaborate offshore in the high seas and deep seabed. The following case study, and other examples included in this chapter, bear testimony to effective regional leadership initiatives which initially require facilitation by a neutral participant but thereafter will develop their own momentum, direction and motivation. Successful maritime clusters are mainly driven by passion and commitment.

CASE STUDY: ARCTIC BUSINESS LEADERSHIP COUNCIL

Background

There are many opportunities for developing business leadership councils. The collaborative process leading to the establishment of the Arctic Business Leadership Council is an excellent example that could be replicated in other regions. The World Ocean Council (WOC) convened a meeting on

16 September 2012 in Reykjavik, Iceland, to discuss the potential for forming an Arctic Business Leadership Council. The meeting brought together more than 30 representatives from a range of ocean and coastal industries.

Context

The WOC is working with its members and the broader ocean business community to catalyze regional cross-sectoral business collaboration in order to foster and coordinate industry input to regional policy processes, science efforts, and the development of solutions to cross-cutting operational and technical issues. In some of these areas, notably the Arctic, there are important government, intergovernmental and NGO efforts underway to address marine and coastal environmental concerns. The relevant industries were not engaged in a systematic, comprehensive, coordinated, cross-sectoral manner for many of these processes related to the Arctic. The WOC is working to engage inputs to, and participation in, specific regional processes to ensure balanced sustainable development, as measured by the regional solutions implemented and the inter-industry and/or stakeholder conflicts reduced. To address this need and opportunity, the WOC initiated an Arctic Working Group from among its members.

The Arctic region is unique with its extreme climate and sensitive ecosystems. At the same time it is rich in natural resources, such as biological resources, energy and mineral deposits. The region is of growing interest for shipping and tourism. The retreat of Arctic ice is providing new opportunities for transport and access to resources. If the Arctic is to be developed in a sustainable manner, responsible companies will play a key role in the economic development of the region and beyond. Sustainable economic growth can stimulate job creation, increase the use of environmentally efficient technologies and provide innovative solutions to development challenges in the region. Cooperation and dialogue among the Arctic states, industry and civil society organizations are critical to this process, which, if done properly, could facilitate and support a unique approach in conducting responsible, sustainable business in the Arctic.

Arctic Business Leadership Council initial meeting

The purpose of the initial meeting was to explore development of an Arctic Business Leadership Council. The meeting was designed to create the possibility for interaction among industry representatives regarding the need, opportunity and value of collaboration and leadership on Arctic issues. The scope of the discussion was primarily sustainability issues within the offshore context, but with consideration of closely related coastal/shoreline issues that have an important influence on the marine areas and issues, notably port development and land-based access to the sea. The approach and setting was a roundtable forum, with company participants providing their perspectives, experience and ideas. There were no formal presentations, except from the WOC to set the stage for the discussion. The focus was on industry representative inputs to allow leadership companies with interests and activities in the Arctic region to shape the structure, process and ongoing work of the newly proposed Arctic Business Leadership Council.

In the preparations for the meetings, several companies identified that the development of an Arctic business leadership group would be valuable because there are many key pan-Arctic and pan-industry issues that could be better addressed though coordinated, collaborative, multi-industry efforts. For example: Arctic Ocean, weather and climate science and observations; transboundary marine ecosystem management and biodiversity conservation; cross-cutting, cumulative impacts (sound, hull fouling/ballast water, invasive species, and marine mammal interactions); marine spatial planning and marine protected area development; and international waters policy and governance.

For most, if not all, of these issues, many companies identified the need and opportunity for standardization of practices and methodologies across the region and the corresponding economies of scale that could be realized through the collaborative research and development of solutions. The potential ongoing work of the Arctic Business Leadership Council could be to engage with the companies that are participating to develop the alignment and agreement on what is being worked on, how the work will proceed, what outputs are most valuable, and how results will be communicated.

Business Dialogue with the Sustainable Development Working Group

The ongoing work of the WOC Arctic Business Leadership Council could also establish the means for regular, coordinated interaction of the private sector with the Arctic Council (AC), the main forum for governmental cooperation in the Arctic region. New business opportunities in the region are one of the main drivers of interest in the Arctic. Despite the rapidly growing interest from businesses, there is no common dialogue forum between the business community and the AC. Four out of the six AC working groups are mandated to deal with climate and environmental challenges.

The Sustainable Development Working Group (SDWG) is the only working group mandated to deal with the human dimension and commercial activities in the Arctic. To discuss the potential mechanism for and the input of the Arctic business community to the SDWG, the Arctic Business Leadership Council meeting was followed by an inaugural Business Dialogue meeting. With the members of the SDWG, this meeting was a first step in developing a forum for the business community to interact with the SDWG and also to consider the interaction with other AC structures and processes.

The Business Dialogue was designed to create an opportunity for the private sector to express its views on Arctic issues, collectively as well as on sector- and geographically specific issues. It allowed industry to raise challenges and opportunities of the Arctic in constructive dialogue with the government representatives, to develop the value of the interchange and participants, and to explore the way forward to regular, sustained interaction between the business community and the AC.

Objectives of the Arctic Business Leadership Council

In the September 2012 meeting, objectives were stated as:

- To bring together company representatives from the key industries in the Arctic, i.e. those currently active as well as those possibly active in the future.
- To identify common areas of challenge and opportunity for industry leadership with respect to sustainable development of the Arctic.
- To explore development of a possible Arctic Business Leadership Council as an instrument to engage the Arctic business community across all sectors.
- To examine ways in which an Arctic Business Leadership Council could foster and facilitate the Arctic business community engaging with other key stakeholders and organizations, especially the AC and its SDWG.
- To develop Arctic business community input to the SDWG meeting.

The potential value of cross-sectoral leadership and collaboration through an Arctic Business Leadership Council was highlighted, especially in the context of sustainable development in the Arctic. The WOC Smart Oceans/Smart Industries program was introduced to create a global system in order to catalyze and coordinate the scaling up of ships and platforms of opportunity to collect and share data. This is a valuable program for areas that are low in data, such as the Arctic. The WOC outlined efforts to develop multi-industry leadership and collaboration at a regional scale with the establishment of an Arctic Working Group. As part of this, a good working relationship was established with the AC and its SDWG through Sweden, which is the current chair of both the AC and the SDWG.

Arctic business: Challenges and opportunities

Many large companies have a commitment to doing sustainable business in the Arctic because business in this region involves long-term strategic entry. A high level of capital and capacity is required to even explore undertaking economic activities in the Arctic: millions is invested in research and development even before starting operations. It therefore follows that companies operating in the Arctic care about their impact and their reputation.

Misperceptions and lack of information can create difficulties in developing and building responsible business operations in the Arctic. The majority of the public has the misconception that the Arctic is pristine, with no industrial activities currently underway in the region. This leads to a need for better communication, both worldwide and with indigenous communities, and the WOC can help to bridge this gap in public perception by 'myth-busting' the misperceptions and underlining the reasonable and responsible use of the Arctic.

In many parts of the Arctic, certain economic activities cannot take place without the participation, if not approval, of indigenous people in the area (e.g. in parts of Canada). Indeed, agreements with indigenous communities are required by law under land-claim agreements. Many companies have extensive interaction with indigenous peoples and ensure their activities are accompanied by efforts to support local communities.

Conclusion

Although the organizational structure of the Arctic Business Leadership Council is in its infancy, there is merit in cross-sector cooperation in support of determination of cumulative impacts, business scenarios and best practice sharing on cross-cutting themes, including adaptive management in response to climate change. However, it will be important to avoid duplication of effort with single-sector industry bodies.

The Arctic Business Leadership Council is exploring a workable set of prioritized issues and activities, and is determining if there is the interest and energy among a critical mass of companies to work on these. Although many sectors and geographic areas have been represented at the early meetings, it will be important to ensure broader participation of key Arctic sectors (e.g. tourism) and countries (e.g. Russia).

Discussion

The Arctic Business Leadership Council is an excellent example of collaboration whereby countries and stakeholders agree to adopt a comprehensive and coherent approach to the maximum sustainable exploitation of the EU's seas and oceans for achieving economic growth.

Another more recent example is Panama and South Africa. The Panama Maritime Authority and South Africa's Ministry of Transport have signed a cooperation agreement in the fields of maritime transportation. The document was signed on 31 July 2015. Both parties have agreed upon a number of areas for cooperation including: safety and security; seafarers and crew matters; marine and maritime training; ports and transportation matters; and recognition of documents. This agreement has important and relevant issues for both parties, such as seafaring training, advice of ports development and management, and those concerning merchant marine safety and security.

Being aware of South Africa's blue economy initiative, which is handled in South Africa by the Department of Transport, Panama is committed to assist with South Africa's know-how and expertise through their Maritime Authority in order to ensure successful implementation of the said African venture. Following this agreement, South Africa will encourage exchange programs in which they will send their youth to be trained as seafarers and receive sea-time onboard Panama-registered ships. The parties agreed to establish a Joint Technical Committee to fast-track the implementation of the agreed areas of cooperation.

References and further reading

African Union (2012). '2050 Africa's integrated maritime strategy'. AU Version 1.0. http://cggrps.org/wp-content/uploads/2050-AIM-Strategy_EN.pdf (accessed: April 2017).

Cicin-Sain, B., Vanderzwaag, D. and Balgos, M.C. (eds) (2015). *Routledge Handbook of National and Regional Ocean Policies*. London and New York: Routledge.

DOSI (2016). 'Deep ocean stewardship initiative'. http://dosi-project.org (accessed: April 2017).

European Commission (2015). 'Summary of the results of the public consultation on international ocean governance'. http://ec.europa.eu/dgs/maritimeaffairs_fisheries/consultations/ocean-governance/doc/ocean-governance-summary_en.pdf (accessed: April 2017).

McGinnis, M.V. (2012). *Ocean Governance: The New Zealand dimension*. Summary Report. Emerging Issues Programme, School of Government, Victoria University of Wellington.

Molinos, J.G., Halpern, B.S., Schoeman, D.S., Brown, C.J., et al. (2015). 'Climate velocity and the future global redistribution of marine biodiversity'. *Nature Climate Change*, 6: 83–88.

Myers, N. and Kent, J. (eds) (2005). *The New Atlas of Planet Management*. Oakland: University of California Press.

Saebo, E. and Cammaert, G. (2011) 'An assessment of international standards for safe exploration, production and transportation of oil and gas in the Barents Sea'. Arctic

Conference, Houston. www.onepetro.org/conference-paper/OTC-22036-MS (accessed: April 2017).

Safina, C. (2005). 'Ocean' in: N. Myers and J. Kent (eds), *The New Atlas of Planet Management*. Oakland: University of California Press.

Tittensor, D.P., Mora, C., Jetz, W., Lotze, H.K., et al. (2010). 'Global patterns and predictors of marine biodiversity across taxa'. *Nature*, 466: 1098–1105.

UNEP (United Nations Environment Programme) (2016). *Regional Oceans Governance: Making regional seas programs, regional fishery bodies and large marine ecosystem mechanisms work better together*. Nairobi, Kenya: UNEP.

UNEP-Nairobi Convention and WIOMSA (Western Indian Ocean Marine Science Association) (2015). *The Regional State of the Coast Report: Western Indian Ocean*. Nairobi, Kenya: UNEP and WIOMSA.

University of York, UK and The Nature Conservancy (TNC) (2010). 'Protecting East Africa's marine and coastal biodiversity: Marine conservation agreements in the Western Indian Ocean'. http://marineworldheritage.unesco.org/wp-content/uploads/2012/01/Protecting-Africas-marine-and-coastal-biodiversity_Marine-Conservation-in-WIO-english.pdf (accessed: April 2017).

UNWTO (United Nations World Tourism Organization) (2013). *Sustainable Tourism Governance and Management in Coastal Areas of Africa*. Madrid, Spain: UNWTO.

World Ocean Council (WOC) (2012). 'Report: Arctic Business Leadership Council meeting'. Reykjavik, Iceland, 16 September 2012.

Chapter 10

The ocean business community

In 1999, Paul Holthus, in his capacity at that time as Executive Director of the Marine Aquarium Council, wrote a ground-breaking article, or perhaps one should say wave-breaking article, on the role of the private sector with reference to the sustainable development of oceans and coasts. The article, 'Sustainable development of oceans and coasts: The role of the private sector', appeared in the journal *UN Natural Resources Forum* and is repeated in this book with permission. Today, Holthus is the Founding President and Chief Executive Officer of the World Ocean Council (WOC). The WOC is the only international, cross-sectoral alliance for private sector leadership and collaboration in corporate ocean responsibility. By joining the WOC, companies and associations worldwide are distinguishing themselves as leaders in ocean sustainability, stewardship and science.

Direct ocean users are industries that depend on the ocean for the extraction or production of goods (e.g. living, non-living and energy) and the provision of services (e.g. transport and tourism). Ocean user support industries are industries that depend on direct users for their existence (e.g. shipbuilders) or drive the need for ocean industry (e.g. extractors, manufacturers, and retailers that transport materials or products by sea). Ocean use infrastructure providers include financial, insurance, legal and other services that enable ocean industries to operate. Without a strong ocean business community, maritime domains could be dominated by public sector authorities with a political agenda. Hence, when dealing with conflicts of interest in the maritime space, there needs to be a balance of public and private sector representation.

'Sustainable development of oceans and coasts: The role of the private sector'

Oceans and coasts provide the majority of the world's ecosystem benefits and much of the private sector is dependent on ocean and coastal resources, services and space. Due to their unique transboundary characteristics, the sustainable development of these areas requires the participation of all stakeholders. However, the international regime for the sustainable development of oceans and coasts is being developed through governmental processes in which the private sector does not have much opportunity for involvement.

Business and industry is recognizing and responding to sustainable development, and ocean and coastal issues are increasingly a part of this. For the private sector to operate sustainably and profitably in ocean and coastal areas, it must understand how it affects these ecosystems and the other users of their resources, services and space, as there is strong public interest in these issues.

Business and industry has the opportunity to develop its role and responsibilities in the sustainable development of oceans and coasts at several levels: a) in relation to consumers; b) at the level of individual corporations; c) at the level of entire business and industry sectors; d) within the private sector as a whole; and e) in relation to the other stakeholders. Important developments are taking place at the first three of these levels, as indicated by numerous examples. Much remains to be done at the important multi-sectoral levels, i.e. within the private sector and in relation to other stakeholders.

A private sector network on oceans and coasts is proposed as a mechanism to ensure information exchange and interaction on the sustainable development of these areas within the private sector. This forum would also provide a vehicle for business and industry input to international processes on oceans and coasts, fostering understanding and interaction between the private sector and other stakeholders.

The private sector and sustainable development

Sustainable development is the responsibility of all parts of society, i.e. governments, public interest groups, consumers, and the private sector (Brundtland 1994). The latter is essential to sustainable development, as business and industry is the principle engine of society – producing goods and services that generate wages, dividends and tax revenue. In developing countries, where much of the world's biodiversity is found, private sector funds from developed countries far outweigh other financial flows (OECD 1995). Over US$260 billion flowed from private sector sources in OECD economies to developing countries in 1995, compared to US$56 billion in official development assistance from OECD governments, which is decreasing. Most of the private sector flow is not to the least developed countries of Asia and Africa that are major repositories of biodiversity.

Economics are becoming central to all conservation issues (McNeely 1997) and it is increasingly evident that economic prosperity and a market that works for the interests of biodiversity are essential to sustainable development (Brenton 1994). As a result, policies need to reflect the use of economic and market instruments, offering government, industry and the public a wider range of options and more effective tools (Sahl and Bernstein 1995).

Although the private sector is focused on its primary role of generating profit, it increasingly recognizes that maintaining economic growth and a healthy environment are inter-related and that the corporate world has a major responsibility (Choucri 1994). This responsibility is embodied in the Business Charter for Sustainable Development of the International Chamber of Commerce and the Principles of the Coalition for Environmentally Responsible Economies (CERES). The corporate signatories to these documents commit themselves to principles and practices of sustainable development, including:

- recognizing the need for long-term economic growth and environmental sustainability;
- developing and providing products and services that have no undue environmental impact, are safe, are efficient in consumption of energy and natural resources, and are able to be reused, recycled or disposed of safely;
- protecting the biosphere, preserving biodiversity and safeguarding habitats;
- informing and working with the public and creating dialogue with stakeholders;
- working towards self-regulation and a demonstrated environmental commitment by corporate management; and
- undertaking environmental audits to measure progress (Schmidheiny 1992; Dias and Begg 1994).

The challenges of sustainable development of oceans and coasts

While the private sector is generally taking note of sustainable development issues and some parts are responding proactively, little of this has focused on oceans and coasts. Ocean and coastal areas provide an enormous portion of the world's ecosystem goods and services upon which the private sector and society depends, e.g. food, recreation, waste-receiving waters, fossil fuels. Most important is the 6% of the world's surface comprising the nearshore marine environment – i.e. estuaries, coastal wetlands, mangroves, coral reefs, and continental shelves. This small portion of the planet provides 43% of the world's ecosystem goods and services, much of which is critical to business and industry (Costanza et al. 1997).

Oceans and coasts also support a significant, unique component of the world's biological diversity, habitats and resources – all of which are connected by a dynamic, three-dimensional water world covering over 70% of the earth's surface (Norse 1993). Due to the fluid nature of the oceans and the international character of coastal ecosystems, their biological and ecological richness and resources often extend over vast geographic scales (Scientific American 1998). However, many of the patterns and processes involved are not well known or understood.

Inadequate understanding of the marine environment has led to the oceans being perceived as having an inexhaustible supply of resources and an unlimited capacity to receive our wastes (*Economist* 1998). This has often been coupled with open access to marine areas and resources and the lack of management for sustainability. Meanwhile, on the land side, population growth and human activities have become heavily concentrated in a narrow band along the coast. The vast majority of major urban centers around the world are located on the coast and all of them are expanding rapidly. In SE Asia 65% of all major cities are coastal, and in Latin America the rate is 75%. Overall, about 37% of the world's inhabitants – over 2 billion people – live within 100 km of the sea, and almost 50% of the world's population live and work within 200 km of the shore (Cohen 1998).

It is now clear that the complex, multi-sectoral mix of human activities on and adjacent to the sea is seriously affecting oceans and coasts. These have led to the unsustainable use of resources, the degradation and destruction of habitat, and extensive pollution of ocean and coastal areas, especially from land-based activities. These impacts are rapidly expanding in scope and increasingly extend across borders and throughout the global "commons" of the seas (Economist 1998). The cumulative and synergistic effects of these impacts complicate our ability to understand and manage oceans and coasts for sustainability. The sustainable development of oceans and coasts is thus inherently multi-sectoral and international, creating unique challenges and requiring the participation of all stakeholders, including business and industry.

The value of oceans and coasts to business and industry

Oceans and coastal ecosystem benefits are critical to the private sector and a substantial proportion of business and industry is entirely dependent upon these resources, services and space. This includes marine transport, port development and operations, shipbuilding, fisheries and fish processing, marine and coastal tourism and recreation, offshore oil and gas, coastal mining, and coastal aquaculture (Clark 1996).

A much wider range of business and industry influences oceans and coasts less directly, predominantly through hydrological linkages in coastal watersheds (Grigalunas and Congar 1995). The most common example of this is through the use of fresh water and/or the discharge of effluent that ends up in nearshore waters. Many less obviously marine-related industries are also sited near the coast, e.g. to use coastal waters for processing or cooling. There are also more encompassing economic linkages. For example, cost-effective marine transport is the basis for many industries and population centers being located at the coast. Additional businesses and industries concentrate in the same areas in order to access and service the coast-dependent industries, marine transport networks, and coastal urban centers.

The result is that the economies of most countries are heavily dependent on ocean and coastal benefits, as illustrated by the US, where one-third of the GDP is produced in coastal areas through fishing, transportation, recreation, and related industries and one of every 6 jobs is marine related (Daley 1998). When broken down by sector, ocean and coastal connections are predominant in many components of the US economy. In the tourism sector, coastal states account for 85% of tourism, with 180 million people spending about US$75 billion on coastal recreation (White 1997; Houston 1996). In trade and transport, marine transportation dominates foreign commerce, handling 2 billion tons of cargo each year, including 95% of US foreign trade. This ship-borne trade passes through 190 US seaports and has a value of over US$500 billion/year. In the fisheries sector, US commercial fisheries generate another US$20 billion/year and global fisheries, including processing and marketing, employs 200 million people (Economist 1998).

In US coastal and island states, ocean and coastal economic links are particularly important. For example, in California, industries that depend directly on a healthy coastal and marine environment generate US$17.3 billion/year and employ 370,000 people. In Hawaii, commercial fishing, ocean recreation, maritime transportation and repair, ocean research, seafood marketing, and aquaculture generated US$2.9 billion in revenues in 1992. These industries are projected to generate US$3.8 billion in revenues and employ more than 20,000 people in 1998, adding about 2,300 new jobs since 1992 (Hawaii DBEDT). Florida's 220 mile long reef tract generates US$1 billion in annual fishing and tourism revenues (Lee 1996). Similar figures are available in Australia where the Great Barrier Reef attracts more than 2 million visitors per year and is worth more than AUS$1 billion annually (Alcock 1996).

Although it is difficult to obtain figures quantifying the value of oceans and coasts to the economies of developing countries, it is evident that the amounts are significant. In SE Asia, for example, almost 90% of fisheries come from shallow continental shelf and coral reefs (Chua and Garces 2012) and fish supply an estimated 65–70% of animal protein in these countries (Valencia 1990; ADB 1993). One km^2 of healthy coral reef produces about 37 metric tons of fish/km^2 (Alcala 1988) and is worth about US$12,000/year in fisheries production and US$1.16 million over 25 years in fisheries, tourism, and coastal protection value (Cesar et al. 1997).

Travel and tourism is particularly important to developing countries and is largely focused on the coastal and marine environment. Much of this is in countries and communities with few economic development alternatives, especially in small island states. In the Caribbean, up to 30% of investment, GDP and jobs are generated by tourism – with 100 million visitors joining the region's 170 million residents each year (Jenner and Smith 1992). In Jamaica alone, tourism and recreation generates US$80 million per year (as compared to fisheries which generate US$25 million/year), with tourists indicating in a World Bank survey a willingness to pay US$387 per person per year to protect coral reefs and the environment (Huber and Ruitenbeek 1997). Coastal areas also account for a significant percentage of tourism in continental developing countries. For example, in Costa Rica and Mexico, beaches are the primary destination for 30–37% of tourists. In Equador, the Galapagos Islands are the major visitor destination (Jenner and Smith 1992).

Competition, conflict and public perception

Given the level of economic dependence on ocean and coastal resources, services and space – and the level of human population that coastal areas support – it is not surprising that there is a high degree of competition. Business and industry is competing with other sectors of society for these amenities and often finds itself in conflict with the government or public (Tol et al. 1996; Turner 1997). There is also a high degree of competition and conflict within the private sector itself, i.e. among the various components that have overlapping needs for ocean and coastal resources, services or space (Hodgson and Dixon, 1988).

The value put on ocean and coasts by the public is a key consideration. A 1996 survey in the US by a leading market survey group found that 87% of people felt that the condition of the ocean is "very important" or "somewhat important" to them (SeaWeb 1996). In coastal communities, fully two-thirds consider the condition of the ocean to be "very important". Over 80% of the US public believes that oceans are threatened by human activity, with over half seeing this as a serious threat to the quality of life. A majority felt strongly about the following ocean issues: oil entering the ocean, overfishing, loss of species, beaches closed due to pollution, and loss of coastal habitat.

In spite of efforts by some components of business and industry to address marine environmental issues, the US public believes that the private sector in general is largely responsible for ocean problems, e.g. that corporations were the major polluters. Oil spills, chemical runoff from "corporate farms", polluted water discharge, and seafood contamination were the most serious ocean problems identified by the public. Of these, people were most "angry" about oil spills and they felt oil companies were the most able to affect the oceans.

In order for the private sector to operate sustainably and profitably in ocean and coastal areas, it must not only understand how it affects the resources, services and space it uses, it must also understand how it affects the other users of those amenities, particularly the public. There is clearly a need and opportunity for business and industry to address ocean and coastal environmental issues and communicate its efforts to the public. Those that respond to this will connect with a large constituency. Conversely, it is also important to understand how the other users of ocean and coastal areas affect the potential for business and industry to operate sustainably. This includes consideration of how the various components of the private sector affect each other, e.g. the impact of oil operations on local commercial fishing. Finally, added to all of these interactions are the effects of the environment itself on ocean and coastal ecosystems, e.g. through climate events such as hurricanes and El Nino.

[. . .]

The role of the private sector in the sustainable development of oceans and coasts

The role and responsibilities of business and industry in relation to the sustainable development in ocean and coastal areas can be considered at five levels:

- consumers;
- individual corporations;
- business and industry sectors;
- the private sector as a whole; and
- the full range of ocean and coastal stakeholders.

Important developments are taking place at the first three of these levels [. . .].

[However, Holthus reflected that it was beyond the scope of his paper] to provide a thorough survey of private sector efforts in relation to the sustainable development of oceans and coasts and more work needs to be done to assemble this information. However, it is clear that at the essential multi-sectoral levels, i.e. within the private sector as a whole and in relation to other stakeholders, the sustainable development of oceans and coasts has not been actively addressed by business and industry and important challenges and opportunities lie ahead.

[. . .]

The characteristics of oceans and coasts and human impacts on them mean that their sustainable development must be addressed through a "critical mass" of effort by those affecting the shared marine and coastal environment.

Business incubators: An introduction

A business incubator is a business that provides advice, equipment, temporary premises, or other facilities to those starting up a business and lacking in capital. It is established to develop and implement a strong microenterprise development program that addresses the needs of the low- to moderate-income residents and provides access to loan capital, business support, training, technical assistance and follow-up services. An incubator can also work with existing small business owners to strengthen and sustain this sector. Most cities recognize the potential positive and significant impact that the development of micro- and small business enterprises has on economic development.

Operating under the office of a development agency, a business incubator could provide services such as training workshops, technical assistance, business support, advice on the loan application and approval process, as well as on loan disbursements and collections, and ongoing business monitoring. For a small business, an incubator could offer services such as marketing, bookkeeping, business coaching, office/conference space for client meetings, and office equipment that would otherwise not be affordable. Non-clients also have the opportunity to take advantage of a business incubator, in a virtual way. Training is offered to incubator businesses and onsite staff will be available to assist.

A job placement centre is part of the business incubator, and it assesses the existing community job skills, potential job opportunities, and skills training

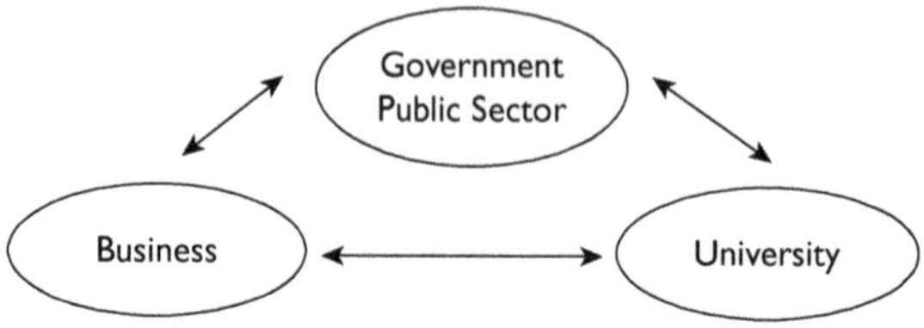

Figure 10.1 Triple Helix linkage model

which is being provided in the town, city or region. An incubator acts as a clearing-house of information, as well as providing assessment, training and placement. Many 'would-be entrepreneurs' do not have the technical know-how to turn their dreams into reality, or how to navigate the bureaucracy involved in opening and running a successful business in their country. A business incubator can provide ideas, start-up space and technical support to new entrepreneurs, as well as established businesses. There are four main types of business incubators: local economic development incubators, academic and scientific incubators, corporate incubators, and private investors' incubators.

The viability and success of business incubators depends largely on the implementation of the Triple Helix model. The Triple Helix model (see Figure 10.1) comprises three basic elements:

- A more prominent role for the university in innovation, on par with industry and government, in a knowledge-based society.
- A movement toward collaborative relationships among the three major institutional spheres, in which innovation policy is increasingly an outcome of interactions among the spheres rather than a prescription from government or an internal development within industry.
- In addition to fulfilling their traditional functions, each institutional sphere also 'takes the role of the other', operating on a *y* axis of their new role as well as an *x* axis of their traditional function. Functional integration, as well as differentiation among institutions, takes place though interaction among the spheres.

Maritime cluster incubation

Kaczynski and Looney (2012) paved the way for the establishment of the first maritime business incubator in South Africa. International experience indicates that the level of the business formation tends to be higher in clusters. Start-ups are more reliant on external suppliers and partners, all of which they find in a cluster. Clusters therefore reduce the costs of failure, as entrepreneurs can fall back on local employment opportunities in the many other companies in the same field. Thirdly, companies and research institutions can achieve higher levels of innovation. Knowledge spillovers and the close interaction with customers and other companies create more new ideas and provide intense pressure to innovate while the cluster environment lowers the cost of experimenting. Maritime clusters can play a key role in developing informational and educational activities to assist the citizenry in dealing with the complex marine resource issues facing South Africa and developing countries. The concept of maritime incubators fits well with a key strategic objective of the Operation Phakisa initiative in South Africa, i.e. to unlock the job creation potential of the ocean economy.

Small business incubation centres

Small business incubation centres have a long tradition in the US, and they are incrementally used by international donors and local funding sources to promote private sector and micro-enterprises in the developing countries. In South Africa business incubators are greatly needed and could be established by a partnership between city councils, chambers of commerce, local township development committees, and supported by a local university. This partnership could be financed by South African funding sources as well as by foreign donors. Improving the welfare of coastal poor and the development of small-scale and micro-enterprises in the coastal zone is one of the key objectives of inter-university collaboration.

For example, collaboration between the University of Washington (UW), Seattle, and Nelson Mandela Metropolitan University (NMMU), Port Elizabeth, stipulates, amongst others, the objective: 'to improve the economic opportunities of the coastal population'. Creating business incubators in South Africa will, therefore, make a valuable contribution to this stated goal, while at the same time meeting pressing needs of the coastal economy and coastal zone population. Small business incubation centres have good potential to assure a quicker and a more realistic way to promote and expand economic development opportunities for coastal communities and to improve the welfare of the poor segments of their population. The incubator model meets most recommendations and expectations expressed in various studies on the past performance and future objectives of various private sector promotion programs.

Obstacles facing small maritime business in developing countries

It is the observation of Kaczynski and Looney (2012) that many would-be entrepreneurs in the coastal areas in South Africa, and developing countries, find themselves faced with almost insurmountable obstacles in converting otherwise well-thought business concepts into actual opportunities for owning and operating successful small coastal businesses and micro-enterprises. In industrialized countries, such as those in the EU, and the US and Japan, an individual wishing to start a new business would normally have advisors or consultants specializing in such disciplines as accounting and finance, law, production and marketing, and human resource management, not to mention various role models and mentors. Furthermore, these specialists and consultants are normally easily accessible at reasonable cost to assist and counsel in the critical early stages of business structuring and business plan writing. In addition, this exercise of early stage business planning and documenting, besides having been shown to dramatically increase the probability of success, is a cornerstone in securing sufficient start-up capital.

American experience in incubation

A tool to promote success in start-up businesses that was applied in the US is that of running comprehensive small business incubation centres. The concept of an incubator is based upon an understanding of the inherent weaknesses and mistakes that early stage entrepreneurs invariably make, and through this understanding, a support incubator centre is established to mitigate these weaknesses and thus maximize the probability of success. In other words, the incubation centre maximizes the probability of success by compensating for, or eliminating, the factors that lead to failure. Through an incubator, real people start real commercial activity, hire real employees, grow a real business, and graduate by moving their business on to a private commercial activity and location.

In the current environment of South Africa, the business climate is ripe for competent would-be small businessmen or businesswomen with the drive and determination to be successful, and the marketplace is hungry for food, new goods and services. A business incubator is not purely a training centre to establish primary occupational skills. Rather, it is a support entity for those with available skills such as tourism services, artisanal fishing and marine aquaculture, fishing/aquaculture and agricultural gear producers and repairers, small-scale marketers, small-scale fish and agricultural product processors, welders, electronic communication and publicity technicians, mechanical establishments, aquaculture consultants, computer servicing specialists, office equipment repair specialist, marine biologists, marine equipment distributors, and other small-scale entrepreneurs.

Founding consortium – The secret of successful incubation

The concept is simple: those with a good reputation (as screened by local community groups), plus some professional experience and skills, are given an opportunity to start a real business with easily attainable support so that they eventually grow in size and develop business skills to the point where they are pushed from the nest (incubator) to make room for the next business to be incubated. The secret to successfully establishing such an incubator is one of creating a founding consortium of the right participants who in total possess and represent the necessary professional disciplines along with the determination and commitment to make the incubator a success. Ideally, as a minimum, this initial founding consortium should include:

- A representative of the local municipality or provincial government (preferably from the department with responsibility for private sector growth), and/or a lending institution (possibly a bank, national foundation, or international donor) with a commitment to socio-economic development.
- One or more experienced private business persons, and/or a senior officer from the local chamber of commerce.
- A representative from an institution that provides primary business training in areas such as accounting, production management, marketing and finance.
- A representative from a lending institution with funds dedicated to economic growth in the developing country – a critical component of the consortium, and, by default, a member with the power of veto.

Without a captive lending institution, the dynamics of the incubator shift away from maturing successful enterprises to one of focusing on raising and maintaining operating capital. In essence, acceptance into the incubator would include, if necessary for start-up, a commitment to provide capital by the lending institution.

It would be the responsibility of the founding consortium to establish the governing mechanics of decision making, as well as oversight of the day-to-day activities of the incubator. To maximize the benefits from the incubator, setting the criteria in the context of South Africa, and also the maritime nations in Africa, for what types of businesses in, for example, artisanal fisheries and aquaculture development (and perhaps size and location), are needed to support growth in the economy. It is also very important to establish a flawless system with complete durable integrity to ensure the absolute unbiased selection of candidates for the incubator based solely on merit.

Operation of a small business incubation centre

In general terms, the incubation centre could operate as follows:

- The Board of Directors established by the City Council or other regional agency would solicit public comments or other local recommendations and consider input and debate amongst them. It is to arrive at a listing of the selection criteria for perhaps 10 or 20 business profiles where there is an interest amongst the consortium in sponsoring.
- The incubator would then post and circulate this list and solicit applications with the predefined criteria for selection clearly identified. From these applications, a systematic and unbiased process would begin to narrow the applicants, eventually leading to final selection. For obvious reasons, the lending institution would possess an absolute power of veto.
- Once selected, an individual (or a group of partners) would be systematically interviewed and counselled to arrive at a mutually agreed to and defined

program and set of requirements (goals) for them to achieve prior to graduation. This program would contain, among other things, a timeline of goals that must be achieved in order to remain in residency and to subsequently graduate.

- From the program would emerge criteria and time constraints for completing a comprehensive business plan, establishing a budget, completing loan and business creation documentation and requirements, acquiring inventory and equipment if needed, a move in date to the incubator, and execution of the comprehensive business plan.
- A full-time support staff at the incubator would then provide training and support as determined necessary from their program and subsequently as determined necessary by direct interaction. This support staff would monitor, and in the early stages actually provide, business administration assistance in accounting, payroll, inventory control, and business record maintenance, eventually transitioning these responsibilities as the business plan dictates, and as the lending institution concurs.
- Utilizing the structure of the incubator, these individual businesses would in fact go about their normal day-to-day activities consistent with building viable and profitable businesses. The professional staff of the incubator would offer support, encouragement and technical assistance as necessary.

Siting for a small business incubation centre

It is recommended that the project leaders determine adequate facility where the incubator could be established. It must be large enough to house perhaps six to eight separate and individual maritime businesses at one time. In addition this facility must have space set up to accommodate business support sections to include:

- A group of expert engineers and technicians.
- Space for accounting and legal counselling.
- General administrative services and business record maintenance (e.g. telephone, fax, copy machine, computers with access to internet, printers, etc.).
- A founding and financing consortium office.
- Location considerations, such as ease of access, transportation, communications, utilities and distance from the local marketing centres.

If the incubator is designed to help micro-entrepreneurs operating close or inside of popular marketplaces or at the beachside, then the design and overall concept of the incubator's location must be properly adjusted. In many cases, incubated enterprises must be located very close to coastal waters, ports and beaches. If a used facility is identified then it usually would need general maintenance, adaptation and repair. In addition, to support a business-incubating program, the property will need additional improvements to accommodate the individuality of the specific businesses.

It is difficult to estimate what these costs might be and the first step must be to retain space planning opportunities and estimates of costs to renovate the facility. It is assumed that donor funding from national and foreign sources could be relatively easily obtained for this innovative and professionally well-designed project.

Professional training of incubating specialists

Based on an extensive experience in designing, negotiating, presenting, implementing and monitoring incubators in the US and some developing countries, it is possible to apply this knowhow to developing maritime nations and their coastal zone's economic development, with particular focus on the small-scale business expansion in coastal communities. An important part of this contribution will be short-term intensive local and overseas training of local incubating specialists (accountants, financial experts, administrators, communications specialists, enterprise management advisors, marketing specialists, business, legal and tax consultants, aquatic food quality and sanitary specialists, marine aquaculture and aquaculture extension workers, and others). Initial expertise in running an incubator could be provided by expatriates, but in the shortest time possible they should be replaced by local specialists. These should first be trained and gain some experience overseas through study tours and intensive short-term professional courses with practical visits to functioning incubators.

The eThekwini maritime business incubator concept

The eThekwini Maritime Cluster (EMC) is a non-profit company that was launched in 2009 in KwaZulu-Natal, South Africa. The EMC provides a platform for collaborative engagement between different levels of government, state-owned enterprises and the maritime community to implement programs of common interest that support the growth and improve performance and competiveness of the maritime industry. In May 2015, the EMC hosted an information session for existing businesses and those businesses that aspire to be active in the maritime industry. The sessions were designed to orientate entrepreneurs about the maritime sector and EMC's incubator for maritime entrepreneurs. In tandem with its key objective, 'fostering growth of small- and medium-sized businesses in the industry, and transformation of the industry generally', the EMC's maritime incubator is set to help small- and medium-sized businesses successfully ply their trade in the industry. Through its Enterprise Development Program, an incubator model was developed by the EMC to equip suppliers in the industry by enhancing their technical skills and business acumen (see Figure 10.2).

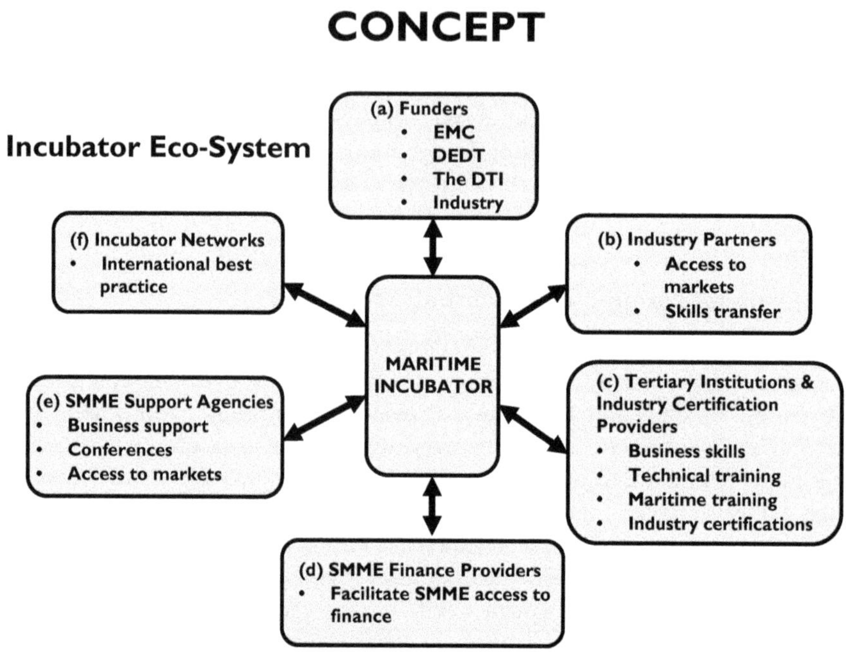

Figure 10.2 EMC maritime business incubator model

(*Source:* eThekwini Maritime Cluster, South Africa, with permission.)

Applicants must meet the following criteria:

- Currently operate a business.
- Have been in business for 2 years or more.
- Employ one person or more.
- Will be willing to learn and make time to attend training and development sessions.
- Have not been on an incubator program since inception of business (but will be considered for the accelerator program).
- Have an enterprise operating in the maritime sector.

CASE STUDY: ENGLAND, LONDON MARITIME BUSINESS ENVIRONMENT

Creating an environment for ocean investment

Ocean investment is attracted to vibrant maritime destinations that offer a wide variety of facilities and services. London and the UK are home to significant numbers of international shipowners, operators and managers as a result of the range of shipping expertise available, a business-friendly tax regime and the advantage of doing business in an international city which is at the heart of the global financial system. London is Europe's largest economic zone. However, other factors are at play to make London attractive. These include: transport links; the ease with which companies can find suitable office locations; the ability to hire highly skilled and expert staff, including experienced ex-seafaring officers working in a range of shipping sectors; and the ability to exploit London as a cosmopolitan capital with world-class schools, universities, hotels, restaurants, theatres, conference venues, and the list goes on. London has become the natural home for shipping and maritime business services. Greenwich Mean Time provides investors with the perfect opportunity to manage their global business, and the UK's geographic location, as a bridge between European, American and Asian markets, makes it the perfect location to run complex worldwide operations. For centuries, the UK has led the way in developing commercial law, and providing the framework for global trade in goods and services and for maritime activity. That legacy continues today, and is stronger than ever.

Benefiting from a highly competitive corporation tax regime, London is the recognized and most comprehensive base for world-leading maritime law, arbitration, insurance, classification, shipbroking and financial services. It also hosts the International Maritime Organization (IMO), the UN's regulator of the international industry. The maritime services sector has seen higher productivity than many other sectors. In the period 2009–2011, 9,500 extra jobs were created (an increase of 6% while the national average fell by 0.4%). In 2011, the combined maritime services sector was almost 10% more productive than the economy-wide average, with ports and business services especially high; the economic cycle had a particular impact in 2009 when the sector was 30% more productive.

As a result, London sees more than 10,000 of the most highly trained, experienced experts in maritime business working across all disciplines, providing world-class services to every corner of the globe. Businesses invest in London because they respect its history and trust its knowledge, expertise and thought-leadership.

London: One-stop shop

London is often called a 'one-stop shop' for maritime business services. The city is home to dozens of companies providing expertise in maritime law, arbitration, classification, shipbroking, insurance, ship management and financial services (see Figure 10.3). Companies benefit from one another, striving to improve in a competitive market, creating a 'race to the top'

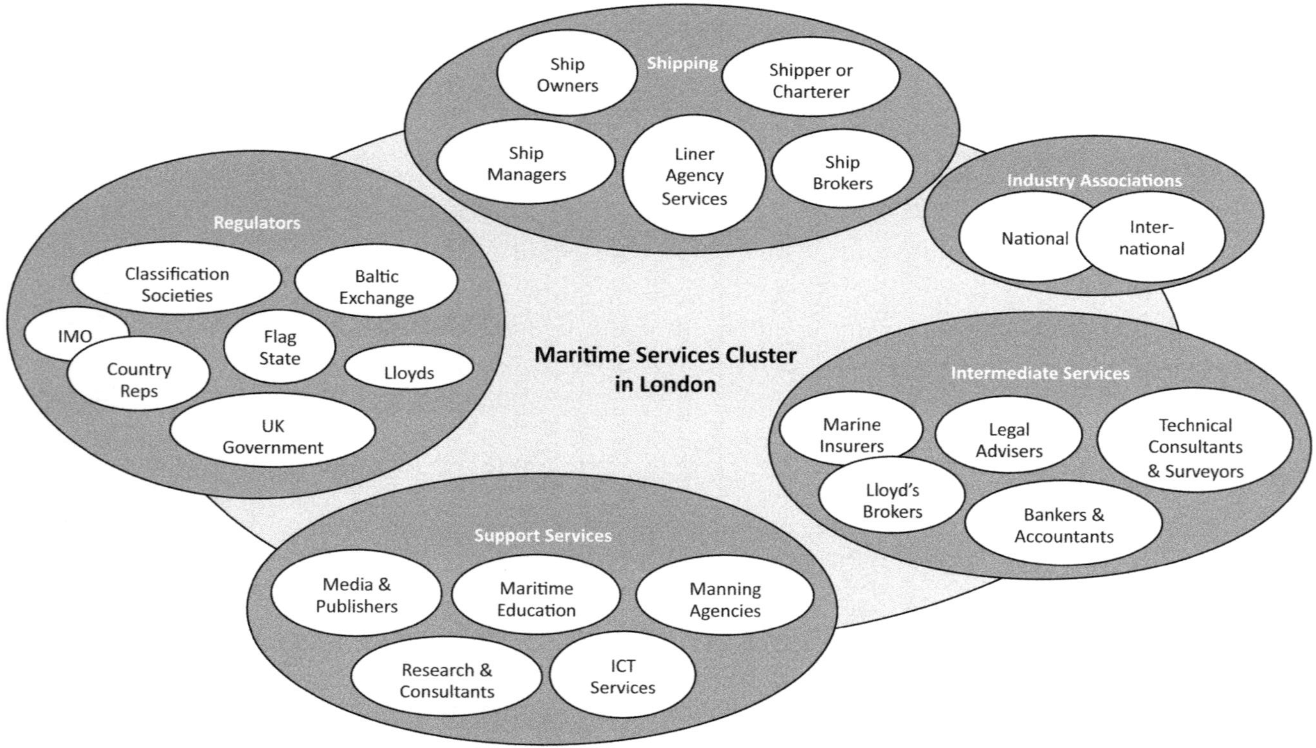

Figure 10.3 Maritime services cluster in London

(*Source:* Maritime Services Ltd London, with permission.)

culture that demands, and achieves, excellence. No other country in the world can boast the breadth and depth of business expertise.

Maritime skills

The UK is globally recognized as a world-class centre for maritime education and training. With internationally recognized educational provision and professional skills development in well-renowned universities and colleges, it meets the varied needs of the global shipping industry. Underpinned by a strong, innovative and highly regarded research capability across a range of disciplines, the specialist maritime departments in a number of universities offer first-class facilities and teaching to UK students and those from overseas. Twenty-six UK universities and colleges offer marine and maritime courses to varying levels, including higher level education, as well as numerous private education centres and professional organizations. There are well over 10,000 students. For school and college leavers, a full range of MCA-approved maritime training is offered. A number of career paths are available, covering training for cadets and ratings, as well as courses tailored for specific needs. Other encouragement for employers includes the Support for Maritime Training budget, Maritime Apprenticeships and the Tonnage Tax Training Commitment.

Marine manufacturing and engineering

Marine manufacturers design and produce high-quality marine equipment products and services to defence, energy, shipbuilding and boatbuilding customers in the UK and across the world. The three broad business segments in which UK marine companies sell their capability are commercial and naval, leisure marine, and energy. UK marine exports totalled £1.4 billion in 2011, exports to non- EU countries have increased five-fold since 2007. Commercial and naval sector strengths include: naval architecture, ship maintenance, repair and conversion, systems integration, navigation, electronics, propulsion, data management, water treatment systems, marine science, and classification societies.

Leisure marine encompasses super-yacht, powerboat and yacht manufacture, marine equipment, small workboats and clothing. UK strengths include: design and bespoke manufacture of leisure boats and rigid inflatables, composite processing, navigation equipment, electronics, deck gear, and rigging equipment. In terms of energy, the UK oil and gas supply chain has a value of around £19 billion, a third of which comes from exports of UK oil and gas. UK companies offer the experience and knowledge of more than 40 years of operating in the North Sea. In relation to offshore wind, the

UK is forecast to invest up to £75 billion in offshore wind farms by 2020 and the growth of its wave and tidal industry could attract up to £4 billion per annum of investment by 2050. UK marine energy capability encompasses fields including: workboats and associated equipment, oceanography services, oil spill detection and environmental services, marine surveying, and financial, legal and insurance services.

The UK also possesses strong research institutions and universities, including marine research and development centres at the University of Southampton, the University of Strathclyde, Portsmouth University and Plymouth University. There are currently more than 150,000 students in the UK studying engineering-related degrees. The UK has a strong breadth of marine capability delivered through a diverse range of channels. A well-established skills base supports the marine sector across the supply chain, particularly in advanced engineering composites and design. In addition to some of the world's leading boat manufacturers and shipbuilders, the UK is home to many companies supplying equipment and other services to the sector. Various associations and organizations represent the views of the marine sector. The UK has many centres of excellence, focusing on various aspects of the marine sector, including the European Marine Energy Centre (EMEC). Scotland is a world-leading region for developing and commercializing the marine energy market.

Conclusion

A little over 30 years ago, port wastelands began to appear, mainly as a result of the relocation of port activities due to the shift to container traffic. These port wastelands were considered opportunities for the city, and the trend was viewed as a type of re-conquest. Today, more than a quarter century after the creation of the Association Internationale Villes Ports (AIVP) in 1988, there is still much debate about the future of port wastelands. Although new port facilities are still being developed outside of the city, we have also witnessed the arrival of a new vocabulary and new strategies: no longer simply the urban re-conquest of abandoned port spaces, but also the survival of active ports in the city; no longer a sharp dividing line, but a balanced mix and an integration of functions. Through AIVP's worldwide network and their frequent exchange of experiences, we have witnessed, in the last few years, the return of the port to the city.

The 14th World Conference Cities and Ports, held in Durban, South Africa in 2014, strongly confirmed that in port cities, faced with global and local challenges, the time has come for partnerships and for shared use of resources and territories. Modern technological developments make it possible to conceive operational solutions that promote a positive industrial ecology in the city/port territories, as well as improved mobility and flows through the interfaces, optimisation of renewable energies, and real integration of port dynamics into the life of the city and the community. The smart port city, which takes into account the concepts of 'connected objects' and 'big data' in the design of projects and sustainable development strategies, is now a reality in several port cities.

Plan the City with the Port: Guide of Good Practices (www.aivp.org) was compiled to provide policymakers and stakeholders with a decision-support tool for addressing the problems they will face when putting into practice the ideal of 'Planning the city with the port'. The guidelines and examples provided do not claim to be complete or exhaustive, but are meant as sources of inspiration to address four major topics: spatial organisation; economic development strategies; environmental challenges; and project management and governance. The guidelines are the result of the intense exchange of experiences that has taken place during the international meetings organised by AIVP, and of the AIVP's constant monitoring of the implementation of development projects at the city/port interface.

Discussion

In the keynote interview at the Sovereign Wealth Fund Institute Summit (SWFI) in Singapore during April 2016, the CEO of the WOC, Paul Holthus, described the opportunities for investing in responsible ocean economic development. He provided an overview of the ocean economy and recent trends for wealth fund operators and asset managers, and he highlighted the opportunities for long-term investment in key sectors, including port infrastructure development and adaptation, sustainable fisheries and aquaculture, offshore renewable energy, and other areas. The plans for the WOC Ocean Investment Platform were outlined to the SWFI summit participants. Following on from the Sustainable Ocean Summit (SOS) 2015 session on 'Investment and innovation for ocean sustainable development', this platform will provide a global structure and process to bring the investment community together with major ocean use companies and the companies providing the technology and solutions for ocean sustainable development challenges.

The WOC Ocean Investment Platform will be a forum to identify, articulate and evaluate ocean industry opportunities for investment in responsible ocean development. Through this, and in collaboration with other forward-thinking platforms such as the SWFI, the WOC will facilitate synergies and economies of scale among investors and the ocean business community as a way to accelerate ocean sustainable development.

References and further reading

ADB (Asian Development Bank) (1993). *Fisheries Sector Profile of the Philippines*. Manila: ABD.

Alcala, A. (1988). 'Effects of marine reserves on coral fish abundances and yields of Philippines coral reefs'. *Ambio*, 17: 194–199.

Alcock, D. (1996). 'Tourism: The key player in the ecologically sustainable development of the Great Barrier Reef'. Paper presented at the 1996 World Congress on Coastal and Marine Tourism, Honolulu.

Brenton, T. (1994). *The Greening of Machiavelli: The evolution of international environmental politics*. London: Earthscan/Royal Institute of International Affairs.

Brundtland, G.H. (1994). 'The challenge of sustainable production and consumption patterns'. *Natural Resources Forum*, 18(4): 243–246.

Cesar, H., Lundin, C., Bettencourt, S. and Dixon, J. (1997). 'Indonesian coral reefs – An economic analysis of a precious but threatened resource'. *Ambio*, 26(6): 345–350.

Choucri, N. (1994). 'Technology and finance for sustainable development'. Prepared for the Environment and National Resources Group of the United Nations Development Program, 17 February 1994.

Chua, T.E. and Garces, L.R. (2012). 'Marine living resources management in the ASEAN region: Lessons learned and the integrated management approach' in: A. Sasekumar, N. Marshall and D.J. Macintosh (eds), *Ecology and Conservation of Southeast Asian Marine and Freshwater Environments including Wetlands*. New York: Springer.

Clark, J.R. (1996). *Coastal Zone Management Handbook*. New York: CRC Press.

Cohen, J.E. (1998). 'Estimates of coastal populations'. *Science*, 278(5341): 1211–1212.

Costanza, R., d'Arge, R., de Groot, R., Farber, S., et al. (1997). 'The value of the world's ecosystem services and natural capital'. *Nature*, 387: 253–260.

Daley, W.M. (1998). 'Ocean update' on seaweb.org, now incorporated into www.maritime.ihs.com (accessed: May 2017).

Department of Environment, Water and Natural Resources (2013). 'Coastal planning information package: A guide to coastal development, assessment and planning policy'. Government of South Australia. www.environment.sa.gov.au/files/sharedassets/public/coasts/coastal-planning-information-package-gen.pdf (accessed: April 2017).

Dias, A.K. and Begg, M. (1994). 'Environmental policy for sustainable development of natural resources: Mechanisms for implementation and enforcement'. *Natural Resources Forum*, 18(4): 275–286.

Economist, The (1998). 'A survey of the sea: The deep green sea'. London. www.economist.com/node/371986 (accessed: April 2017).

Grigalunas, T. and Congar, R. (eds) (1995). 'Environmental economics for integrated coastal area management: Valuation methods and policy instruments'. UNEP Regional Seas Reports and Studies No.164. Nairobi: UNEP.

Hawaii DBEDT (Department of Business, Economic Development, and Tourism). http://dbedt.hawaii.gov/ (accessed: April 2017).

Hodgson, G. and Dixon, J. (1988). 'Logging versus tourism and fisheries in Palawan: An environmental and economic analysis'. Occasional Paper No.7. Honolulu: Environment and Policy Institute, East West Center.

Holthus, P. (1998). 'Workshop on incentives private sector partnerships and the marine and coastal environment'. Report of the 8th Global Biodiversity Forum, Montreal,

August 1997. Gland, Switzerland: International Union for Conservation of Nature (IUCN).
Holthus, P. (1999). 'Sustainable development of oceans and coasts: The role of the private sector'. *UN Natural Resources Forum*, 23(2): 169–176.
Houston, J. (1996). 'The economic value of US beaches'. Paper presented at the 1996 World Congress on Coastal and Marine Tourism, Honolulu.
Huber, R.M. and Ruitenbeek, H.J. (eds) (1997). 'Marine system valuation: An application to coral reef systems in the developing tropics.' Interim Report to World Bank Research Committee, Project #RPO 681-05. World Bank: Washington.
ICS (International Chamber of Shipping) (1997). *Shipping and the Environment: A code of practice*. London: ICS.
Jenner, P. and Smith, C. (1992). 'The tourism industry and the environment'. Special Report No.2453. London: The Economist Intelligence Unit.
Kaczynski, V.M. and Looney S.W. (2012). *The Role of Maritime Clusters in the Incubation of Small Marine Business Enterprises in the Nelson Mandela Bay Area*. Seattle, WA: School of Marine and Environmental Affairs, University of Washington; and Missoula, MT: Line Haul Logistics, Inc.
Lee, D. (1996). 'The economics of managing Florida's coral reefs'. Paper presented at the 1996
World Congress on Coastal and Marine Tourism, Honolulu.
Maritime UK (2011). *Open for Maritime Business*. London: HM Government.
Maritime UK (2015). *Maritime Business Services Strategic Partnership*. London: Department of Transport, HM Government.
Maritime UK (2016). *The UK's Global Maritime Professional Services: Contribution and trends*. London: City of London Corporation.
McNeely, J.A. (1997). *Conservation and the Future: Trends and options towards the year 2025*. Gland, Switzerland: International Union for Conservation of Nature (IUCN).
Norse, E. (1993). *Global Marine Biodiversity: A strategy for building conservation into decision-making*. California: Island Press.
OECD (Organisation for Economic Co-operation and Development) (1995). *Linkages: OECD and developing economies*. Paris: OECD.
Sahl, J. D. and Bernstein, B.B. (1995). 'Developing policy in an uncertain world'. *International Journal of Sustainable Development and World Ecology*, 2: 124–135.
Schmidheiny, S. (ed) (1992). *Changing Course: A global business perspective on development and the environment*. Cambridge, MA: MIT Press.
Schmidheiny, S., Chase, R. and DeSimone, L. (1997). *Signals of Change: Business progress towards sustainable development*. Geneva: World Business Council for Sustainable Development (WBCSD).
Scientific American (1998). 'The oceans'. *Scientific American Quarterly*, 9(3).
SeaWeb (1996). 'Presentation of findings from a nationwide survey and focus groups conducted by the Mellman Group'. Now incorporated into www.maritime.ihs.com (accessed: May 2017).
Stone, D., Ringwood, K. and Vorhies, F. (1997). *Business and Biodiversity: A guide for the private sector*. Switzerland: World Business Council for Sustainable Development (WBCSD) and IUCN–The World Conservation Union.
Tol, R., Klein, R., Jansen, H. and Verbruggen, H. (1996). 'Some economic considerations on the importance of proactive integrated coastal zone management'. *Ocean and Coastal Management*, 32(1): 39–55.

Turner, R. (1997). 'Ecological economics and project management in the coastal zone'. *Terramare*, 1: 28–29.

Valencia, M. (1990). 'International conflict over marine resources in Southeast Asia: Trends in politicization and militarization' in: L.T. Ghee and M.J. Valencia (eds), *Conflict over Natural Resources in Asia and the Pacific*. United Nations University Press, Oxford University Press and Ateneo University Press, Manila.

White, R. (1997). 'Rethinking integrated coastal management into the next millennium'. Paper presented at Coastal Zone 97, Boston.

Chapter 11

Organized ocean business

The need for a council to represent the ocean business community

Although the ocean covers 71% of the earth's surface, it is an increasingly crowded place. Ocean industries of all kinds are expanding rapidly and impacts on the marine environment are expanding at an ever-increasing rate and global scale affecting endangered species, ecosystem health, critical habitats and coastal communities that depend on marine areas for food and livelihood. Conflicts in the use of ocean space and resources are on the rise among industries, and with other ocean stakeholders. Ocean industries are being held accountable for their impacts by government, intergovernmental and non-governmental stakeholders who are pursuing increased marine environmental regulation. Private sector access to ocean resources, services and space is increasingly at risk due to the loss of the 'license' to operate and the emerging ocean governance regimes. Unfortunately, there has been little or no interaction among like-minded responsible companies across the sectors to develop leadership and collaboration on ocean sustainability and to engage in systematic, coordinated involvement with other ocean stakeholders.

The private sector is the primary ocean user and is best placed to develop and implement the practices needed to ensure marine ecosystem use is sustainable, and an increasing number of companies and industry associations are tackling the environmental effects of their ocean activities. However, in a global, interconnected ocean 'commons', the actions of one company or even an entire sector are not enough to address cumulative impacts of growing ocean use by a diverse range of industries. There is clearly a need for ocean industries to collaborate within and across sectors to address impacts, reduce conflicts, develop proactive ocean sustainability leadership and constructively engage other stakeholders.

The World Ocean Council

As has been mentioned in previous chapters, the World Ocean Council (WOC) is an unprecedented, innovative approach to addressing ocean sustainability through a global, cross-sectoral industry alliance on 'corporate ocean responsibility'. The WOC is bringing together leaders from a wide range of ocean industries to collaborate in a shared goal of healthy and productive seas and their sustainable use, development and stewardship by a responsible ocean business community. The WOC is the only international organization bringing together the broad range of ocean industries to address global sustainability of the seas by achieving the following objectives:

- A vibrant international, cross-sectoral ocean business community for leadership in sustainability and access to ocean areas and resources by responsible companies.
- An ocean business community well informed on ocean policy processes and constructively engaged with ocean stakeholders.
- Cross-sectoral industry collaboration in practical, cost-effective, science-based solutions to marine environmental issues.
- Improved environmental performance of ocean industries through continuous improvement, best practices and standards.
- Operational ocean industry leadership alliances in priority geographic sea areas.
- Improved mutual understanding among ocean industry sectors, increased cross-sectoral dialogue, and reduced ocean use conflicts.
- Collective industry support for, and participation in, improved ocean science.

The benefits of membership in the WOC include:

- Reduced risks from better tracking and information on ocean policies and decision making.
- Decreased risks due to improved relationships and stakeholders.
- Economic benefits of collaboration in developing solutions to shared, pre-competitive issues.
- Increased participation and coordination on international ocean policy developments.
- Competitive advantage of proactively addressing environmental impacts.
- Increased credibility and recognition of industry actions on ocean sustainability.
- Sharing of pre-competitive best practices in marine environmental responsibility.
- Reduced user conflicts with other industries.

The WOC structure and legal status

The WOC is a business association for ocean sustainability, science and stewardship. Membership is open to dues-paying companies and affiliated organizations that subscribe to the WOC Membership Statement. The WOC working groups will convene around 'platforms' to advance the WOC vision and mission through work on priority issues that the members have identified for inclusion in the WOC work program. The WOC is incorporated in the US and the UK/Europe as a not-for-profit organization, financed primarily through membership fees. Specific projects and programs are financed through donations and grants being sought from governments, international organizations and donors.

Currently, the WOC is a fiscally-sponsored project of the Trust for Conservation Innovation (TCI). The TCI is a non-profit, tax-exempt 501(c) (3) organization under the rules of the US Internal Revenue Service, whose mission is to develop and support projects that seek to transform public and private enterprises into sustainable operations that conserve and protect the environment. Tax-deductible contributions received by the TCI are re-granted to the WOC in support of its mission to catalyze global leadership and collaboration in ocean sustainability and corporate ocean responsibility. The founding members of the WOC constitute a Steering Committee that works with the Executive Director to guide the WOC through its establishment as a separate organization, including developing the governance structure, policies and procedures. The WOC will be developing a stakeholder advisory council, or similar structure. This will create a mechanism enabling ocean stakeholders committed to a constructive dialogue and collaboration on ocean sustainable development to engage with the business and industry representatives participating in the WOC.

Regional Ocean Councils

The WOC encourages the establishment of Regional Ocean Councils (ROC): cross-sectoral business collaborations on marine environmental sustainability that could bring together the range of marine industries in a region.

These primary users of marine space and resources hold the key to future marine ecosystem health and are well placed to implement sustainability solutions that work for business. Priority issues include reducing water pollution, preventing maritime accidents, avoiding the introduction of invasive species, reducing/cleaning up marine debris, and improving marine science and ocean monitoring. A ROC, which could be closely aligned to a Chamber of Business representing organized business in a coastal city, would strengthen a local maritime cluster especially when there is a need for advocacy and lobbying on maritime issues at a high level.

CASE STUDY: UNITED STATES, NORTHEAST REGIONAL OCEAN COUNCIL

Introduction

The Northeast Regional Ocean Council (NROC) is a regional ocean partnership that provides a forum for the New England states and federal agencies in the US to address ocean and coastal issues that benefit from a regional response. NROC was formed to augment current state, multi-state and federal governance and institutional mechanisms to improve management of ocean and coastal resources. NROC encourages and values partnerships and collaboration with other regional organizations and groups to advance shared interests.

Mission

NROC provides a voluntary forum for New England states and federal partners to coordinate and collaborate on regional approaches to support balanced uses and conservation of the Northeast region's ocean and coastal resources.

Scope

NROC was formed in 2005 by the New England Governors and works to:

- Identify and address local, state and regional ecosystem-based environmental issues.
- Seek new, and encourage existing, ocean and coastal initiatives and partnerships.
- Facilitate information exchanges, including reports, discussions, initiatives and plans which may be developed or considered.
- Foster regional communication, interaction and cooperation on marine and oceans-related research and development, education, exploration, and observation.

Structure

Council membership

Each NROC member state – Maine, New Hampshire, Vermont, Massachusetts, Rhode Island and Connecticut – appoints up to two senior level representatives. Governors' appointees may designate an alternate to serve on the Council in their absence. (Alternates are consistent from meeting to meeting.) To maintain continuity, council members remain appointed until removed by a Governor through a reappointment.

Federal agencies are equal members of the Council and are represented by: the Department of Commerce through the National Oceanic and Atmospheric Administration (NOAA); the Department of the Interior through the US Geological Survey, Bureau of Ocean Energy Management, US Fish and Wildlife Service, and National Parks Service; the Department of Agriculture through the USDA Natural Resources Conservation Service; the Department of Defense through the US Army Corps of Engineers; the Department of Homeland Security through the US Coast Guard; and the US Environmental Protection Agency.

One state representative and one federal representative co-chair NROC for an agreed upon period (i.e. 12 or 18 months). These individuals organize agendas, run meetings, track progress on the biannual work plan, and act as principal points of contact and spokesmen for the Council. In addition, they present an annual report to the NE Governors' Conference. State chairs rotate in order based on geography from north to south.

Committees and working groups

The Executive Committee consists of the current NROC co-chairs, as well as the previous and upcoming sets of co-chairs, which amounts to three state and three federal representatives. Examples of Executive Committee duties include attending to administrative matters between meetings and developing recommendations to the Council on projects and partnerships.

Committees are formed to work on Council priority issues and goals, by developing and implementing 2-year work plans. The committees inform and recommend to the Council how best to approach regional problems and to coordinate activities. In addition, working groups are convened to take on specific projects and develop recommendations for approval by the Council. Working groups form as needed and disband upon completion of a project. Each committee and working group has representatives from pertinent state and federal agencies and includes representatives from NGOs, academia and industry. Each committee and working group elects a representative to call and chair meetings.

NROC values and benefits from having close and collaborative relationships with other regional organizations. NROC has Memoranda of Understanding with the Northeast Regional Association of Coastal and Ocean Observing Systems (NERACOOS), the Regional Sea Grant Consortium, and the Gulf of Maine Council on the Marine Environment, and participates in the activities of these organizations. NROC communicates directly with the New England Governors and the President's National Ocean Council and its Steering Committee and Governance Coordinating Committee to make clear the priorities and needs for the Northeast. NROC also works closely and communicates with the Northeast's Regional Planning Body for ocean planning.

Operations

- *Meetings:* NROC meet three to four times a year. Meetings take place in a central location to facilitate participation. The Executive Committee holds regular meetings, every 2 to 3 weeks, via conference call. Committees and working groups meet as necessary.
- *Decisions:* NROC develops recommendations on a consensus basis. If consensus cannot be achieved, a vote is held and the concerns of abstaining or opposing members are made known to appropriate agencies. In a voting situation, there is equal representation between state and federal members, and each member state and federal agency has one vote.
- *Activities:* NROC executes strategies and activities identified in a biannual work plan. NROC produces an annual summary of activities and accomplishments.
- *Support:* NROC seeks funds to support the implementation of its goals. NROC also seeks assistance with staffing and preparing meetings, recording action items and tasks, and other administrative tasks as needed.
- *Communications:* To enhance internal communications and file and work sharing, NROC maintains a Council share site (e.g. Base camp). This site is limited to member state and federal agency Council and Committee representatives; content and access is treated as internal and confidential. NROC also maintains an external public website for broader communication purposes.

Conclusion

Regional ocean partnerships like NROC are recognized in reports issued by the US Commission on Ocean Policy and the Pew Oceans Commission as key mechanisms to enable coordinated, ecosystem-based management approaches to identify regional goals and priorities, to improve responses to regional needs, and to develop and disseminate regionally significant research and information. The Council outlines its issue areas, defines its strategic priorities, and develops and implements 2-year work plans detailing specific strategies and activities. Council meetings are open to all stakeholders and interested members of the public. Meeting agendas specifically include opportunities for partners and stakeholders to provide updates and feedback. Council meetings are held in different locations throughout the region to ensure balanced opportunities for attendance, and meeting notices are posted on the NROC website. NROC continues to be a leader in regional ocean partnerships, demonstrating the benefits of how close coordination and collaboration among state and federal government agencies working together with other regional entities, organizations, interests and stakeholders can develop and implement regional approaches to many of our ocean management challenges.

Discussion

Commercial and recreational fishing, transportation, renewable energy production, telecommunications, science and research, tourism, and recreational interests and industries are just a few of the many ways that people benefit from a vibrant ocean environment. As a region's population increases and ocean uses intensify, ocean regions will be faced with a new generation of ocean management challenges and opportunities. Increasing amounts of plastic and micro-plastic waste pollution continue to have negative impacts on ecosystem function and to limit the quality of life in our coastal communities. Increased activity and disturbances in offshore areas threaten sensitive and unique habitats and ocean features. All these issues and activities require a regional approach.

References and further reading

The Aspen Institute (2013). *The Ocean Community Report: Building an aligned and supportive ocean conservation community*. Washington, D.C.: The Aspen Institute.

Blue Earth Consultants: www.blueearthconsultants.com (accessed: April 2017).

Blue Earth Consultants. 'Ocean stewardship in the private sector'. www.blueearthconsultants.com/wp-content/uploads/2014/07/Blue-Earth-Consultants_Private-Sector_Flyer.pdf (accessed: April 2017).

Chan Kim, W. and Mauborgne, R. (2015). 'Blue ocean strategy: Creating new market space where competition is irrelevant'. INSEAD Blue Ocean Strategy Institute: https://centres.insead.edu/blue-ocean-strategy/ (accessed: April 2017).

European Parliament Committee on Industry, Research and Energy (2015). 'Ocean research in Horizon 2020: The blue growth potential'. Study for the ITRE Committee. www.europarl.europa.eu/RegData/etudes/STUD/2015/518775/IPOL_STU(2015)518775_EN.pdf (accessed: April 2017).

Lainos, I. (2011). *Red Ocean vs Blue Ocean Strategies*. Greece: University of Piraeus.

Mid-Atlantic Regional Council on the Ocean (MARCO) (2016). 'Mid-Atlantic Regional Ocean action plan: Executive summary'. www.boem.gov/OAP-Executive-Summary/ (accessed: April 2017).

Northeast Regional Ocean Council (NROC) (2012). 'Northeast Regional Ocean Council: Terms of reference'. http://northeastoceancouncil.org/wp-content/uploads/2012/10/NROC-Terms-of-Reference-August2012.pdf (accessed: April 2017).

Seo Park, K. (2014). 'A study on rebuilding the classification system of the ocean economy'. National Ocean Economic Program, Monterey Institute of International Studies. http://centerfortheblueeconomy.org/wp-content/uploads/2014/11/10.29.14.park_.kwangseo.the_ocean_economy_classification_systemfinal_21.pdf (accessed: April 2017).

Spalding, M.D., Brumbaugh R.D. and Landis, E. (2016). *Atlas of Ocean Wealth*. Arlington, VA: The Nature Conservancy.

United Nations Environment Programme (UNEP) (2016). *Regional Oceans Governance: Making regional seas programs, regional fishery bodies and large marine ecosystem mechanisms work better together*. UNEP Regional Seas Report and Studies No.197. Nairobi: UNEP.

US National Ocean Council (2013). *Marine Planning Handbook*. US government.

World Ocean Council (WOC): www.oceancouncil.org (accessed: April 2017).

Chapter 12

Tourism in coastal areas

Introduction

The origins of tourism in coastal areas go back to Roman times, when the first villas were constructed in the southern part of the Apennine peninsula. In the centuries that followed, especially from the mid-eighteenth century onwards, coastal tourism was generally related to the therapeutic properties of sea and sun. Sun, sea and sand have continued to provide the main ingredients for coastal tourism until today, especially in the second half of the twentieth century, which was marked by the development of mass tourism.

Coastal tourism is based on a unique resource combination at the interface of land and sea offering amenities such as water, beaches, scenic beauty, rich terrestrial and marine biodiversity, diversified cultural and historic heritage, healthy food and good infrastructure. It includes a diversity of activities that take place in both coastal zone and coastal waters, which involve the development of tourism capacities (e.g. hotels, resorts, second homes and restaurants) and support infrastructure (e.g. ports, marinas, fishing and diving shops, and other facilities).

Coastal recreation activities, which have been increasing both in volume and in number during the last decade, occupy a unique place in coastal tourism. They take in two main types of recreational uses of coastal zones: consumptive and non-consumptive types. Activities such as fishing and shell collection belong in the first category, while activities in the second include swimming, diving, boating, surfing, wind-surfing, jet-skiing, bird watching, snorkeling, etc.

Coastal tourism is strongly dependent upon both natural resources (i.e. climate, landscape and ecosystems), and cultural resources (i.e. historic and cultural heritage, arts and crafts, and traditions). It encompasses activities that can only be carried out in particular areas and in specific conditions. Therefore, certain areas are considered to be particularly suited to specific types of tourism activities, for which they become known on a global scale. Examples include sailing in the Gulf of Mexico, surfing on the beaches of Australia and Hawaii, or scuba diving in the Red Sea.

Besides physical conditions, the development of tourism in coastal areas is related to socio-economic features of the receiving environment such as local community interests, health and security conditions, political factors including unpredictable crises, exchange rate fluctuations, and traditional models of tourism exploitation or, simply, a successful or less effective marketing-led depiction of a destination. Environmental conditions such as unpredictable climate conditions, algae blooms, winds and the associated risk of forest fires, tsunami, storms and floods, as well as many other constant features or unexpected events, also affect tourism development in coastal areas.

The magnitude and economic importance of coastal tourism

Tourism counts among the main economic activities in many countries. Research studies over many years have indicated the popularity of coastal tourism. Statistics from the the World Tourism Organization (UNWTO 2007) show that tourism is the world's largest industry with regard to the number of people involved and the economic profit. According to European data (INRA (EUROPE) 1998), some 63% of European holidaymakers prefer the coast as compared to 25% favouring mountains, 25% preferring cities and 23% choosing the countryside.

Although there is no reliable data on coastal tourism alone, it is generally considered to be one of the fastest growing forms of tourism in recent decades. UNWTO statistics show that 12 of the 15 world's top destination countries in 2000 were countries with coastlines. As an illustration, Bridges (1997) reported that three US coastal states (Florida, California and New York) hosted 74% of a total of 20.6 million overseas visitors to the US in 1995, generating 85% of tourist-related revenues. According to the same author, beach tourism in the US generates US$640 billion annually, which equals 85% of tourist-related revenues (Bridges 1997).

Another example of the magnitude of coastal tourism is the Mediterranean Basin. With all the difficulties of extrapolating data from the statistics collected on Mediterranean countries as a whole, UNWTO estimates that Mediterranean coastal areas alone hosted some 250 million visitors in 2008. As forecasted by Blue Plan (Benoit and Comeau 2005), this number could increase to 312 million by 2025. According to the European Environment Agency (EEA 2005), peak population densities on the Mediterranean coast of France and Spain reach 2,300 people per square kilometre, which is more than double those of the winter season.

The economic importance of coastal tourism is unquestionable. Some figures for the above two regions speak clearly: a study by Wilson and Wheeler (1997) shows that predominantly coastal tourism in California is, with US$9.9 billion, the largest contributor to the state's economy, followed by ports (US$6.0 billion) and offshore oil (US$860 million). EEA data for France – largely water-based rather than land-based (e.g. boating, yachting, cruising and nautical sports), but includes the operation of landside facilities, and manufacturing of equipment, ports and marinas, for instance – shows that tourism provides 43% of jobs in French coastal regions, generating more revenue than fishing or shipping. Similarly, the coastal tourism sector in other regions of the world is increasingly growing in importance with regard to its magnitude and contribution to national economies, as well as to the wellbeing of local communities.

The contribution of coastal tourism to the blue economy

Coastal tourism covers beach-based recreation and tourism (e.g. swimming, surfing and sunbathing), and non-beach related land-based tourism in the coastal area (i.e. all other tourism and recreation activities that take place in the coastal area for which the proximity of the sea is a condition), as well as the supplies and manufacturing industries associated with these activities. Maritime tourism covers water-based activities (e.g. boating, yachting, cruising and nautical sports), but also includes the operation of landslide facilities, manufacturing of equipment, ports and marinas, and services for this segment of tourism.

A comprehensive research study undertaken by Houston (2008), confirmed that revenue generated from US beaches far exceeded any other tourist attractions, including national parks and theme parks. The popularity of beaches dominates tourism. However, in many coastal countries the contribution of coastal tourism is not extracted and expressed as a percentage of total tourism. Hence many coastal authorities do not take coastal tourism seriously and do not dedicate financial and human resources for the maintenance and protection of beaches and coastal zones. Furthermore, coastal management plans are too general and fail to include a risk assessment and safety management plan for each beach and coastal zone. Conditions at each beach and coastal zone are not the same and with the impact of climate change incidents of drowning due to rip currents is increasing.

Coastal and maritime tourism: A source for growth and jobs

The extraordinary beauty, cultural wealth and great diversity of Europe's coasts make them the preferred destination for many holidaymakers from Europe and worldwide. On offer is a wide range of facilities and activities for those who seek recreation for mind and body. Tourism demand continues to increase worldwide. Coastal and maritime tourism is the largest maritime activity in Europe and employs almost 3.2 million people, generating a total of €183 billion in gross value added and representing over one-third of the maritime economy. The challenge for Europe's coastal and maritime resorts is to exploit potential sustainably in order to offer attractive jobs to its people.

Coastal and maritime tourism can be a major source of growth and jobs, especially for the young. The European Commission (2015) has issued a European strategy aiming to bring all stakeholders together to develop innovative and smart solutions to the multiple challenges facing the sector. This strategy sets out 14 targeted actions involving national, regional and industry-level partners.

The challenges for coastal tourism

Tourism is a growing business: in 2013, the number of nights spent in hotels or similar establishments in the EU reached a peak of 2.6 billion. Considering that almost half of these nights are spent by the sea, the coastal tourism industry is faced with huge challenges. These challenges are exacerbated by business fragmentation, difficulties in accessing finance and investment, and by environmental pressures. Average daily expenditure by tourists has been decreasing since the mid-2000s, with a further 9% decrease in annual expenditure expected between 2011 and 2020. A European strategy can help businesses to overcome these difficulties, but destinations should also build on their strengths, enhance cooperation and best practice sharing, and create transregional and transnational partnerships. Together, they can help develop innovative and attractive offerings, in order to appeal to tourists travelling in low season.

Cruising is an important sector of maritime tourism, where marine regions such as the Mediterranean or Baltic can be covered in the course of a week's holiday. The Greek maritime municipality of Marathon is a truly unique destination and a historic symbol. It has invested in high-quality services to provide easy access for people with disabilities. For this initiative, Marathon was selected for the 2013 EDEN award (European Destinations of Excellence).

Opportunities from cruise tourism

The 2012 year-end passenger forecast indicated that a record 20.3 million passengers cruised globally during the year, with 17.2 million sailing from North America, including 11.7 million who live in the US and Canada. Coupled with an annual occupancy percentage that exceeded 102%, this annual passenger growth for 2012 showed continued consumer interest in cruising and an industry where demand continues to outstrip supply. In 2012 alone, 13 new ships debuted from the Florida-Caribbean Cruise Association (FCCA) and/or Cruise Lines International Association (CLIA) Member Lines, presenting new options for guest capacities ranging from 32 to 3,690, and totalling 17,774 at double occupancy.

In 2013, the growth of the cruise industry continued, with an additional 11 innovative, feature-rich ships, international ports-of-call and convenient departures from proximal embarkation cities. The industry's success is headlined by the Caribbean, the dominant cruise destination, accounting for 37.3% of all global itineraries in 2013.

From 2013 to 2016, there were 26 new ships (14 ocean-going vessels and 12 riverboats/coastal vessels) from FCCA and/or CLIA Member Lines with over 48,000 berths and a capital investment value of over US$9.5 billion. Based on known ship additions and deletions, the industry forecasted 21 million passengers to cruise globally in 2013, with 17.6 million of these guests predicted to sail from North America (11.8 million sourced from North America and about one-third

(5.8 million) comprised of international passengers). This represents a 2.5% increase over 2012's figures.

Although the Caribbean continues to have the leading share of cruise industry capacity, there has also been growth in all global cruise regions. In 2012, European ports had 29.3 million passenger visits, a 75% increase over 2006. These arrivals generate wealth and business opportunities but also represent big challenges for the receiving ports. For coastal regions it is not always easy to capture economic benefits generated by cruise tourism, while pressures to invest in port infrastructures and to preserve the environment are increasing.

Cruise ships as a legacy of pollution

Cruise ships are floating cities that produce enormous volumes of completely unregulated or inadequately regulated waste. Today's largest ships can transport more than 5,000 passengers and crew, and have the capacity to generate more than 11 million gallons of wastewater every day, as well as carrying significant amounts of hazardous chemicals from onboard printing, photo processing, and dry cleaning operations. Unfortunately, the cruise industry has a dismal environmental record. A report released by the US General Accounting Office (GAO 2000) found that, from 1993 to 1998 alone, cruise ships were involved in 87 confirmed cases of illegal discharges of oil, garbage and hazardous wastes into US waters, and have paid more than US$30 million in fines. In reality, this number drastically understates the magnitude of the problem, representing only a fraction of the actual illegal dumping being carried out by the cruise industry. In addition to the 87 confirmed cases, there were an additional 17 cases that were referred to the countries where the cruise ships were registered. Some of the 87 cases involved multiple incidents of illegal dumping that, according to the Department of Justice, numbered in the hundreds over the six-year period. Furthermore, this figure represents only the number of detected cases; the aforementioned GAO report reveals that the US Coast Guard's detection and enforcement capabilities are hindered by numerous shortcomings. Foreign-flagged cruise ships represent approximately 1.5% of all foreign-flagged ships entering US ports; however, they are responsible for considerably more than this proportion (4%) of the total confirmed illegal discharges committed by all types of foreign-flagged ships entering US ports during this period.

The cruise industry's repeated violation of environmental laws is a matter of grave concern. In a particularly disturbing case, Royal Caribbean Cruises Ltd. admitted to routinely dumping waste oil from several of its ships, and deliberately dumping hazardous chemicals from photo processing labs, dry cleaning operations and print shops into several US harbours and coastal areas, over a period of several years. Ships were rigged with secret piping systems to bypass pollution treatment equipment. Investigators said the company's violations were so unscrupulous that they characterized the case as a 'fleet-wide conspiracy' by Royal Caribbean to 'use our nation's waterways as its dumping ground,' and so pervasive that the criminal conduct amounted to a routine business practice. The company pled guilty to a total of 21 felony counts in six US jurisdictions, and agreed to pay a record US$18 million in criminal fines (*Los Angeles Times* 1999).

Bluewater Network

Since it began in 1996, Bluewater Network has been a leader in efforts to reduce environmental harm from vessels, vehicles and craft. Bluewater Network created California's historic law curbing greenhouse gas emissions, a law which has already been copied by seven other states and Canada, and it has convinced the US Environmental Protection Agency (EPA) to revise overstated fuel mileage labels on new cars. Bluewater Network's other campaign efforts include successes banning jet-ski use in most of the nation's national parks, cleaning up pollution from cruise ships and commercial vessels, and leading the effort to introduce biofuels into the nation's fuel supplies in order to combat global warming.

Launched in 1969, Friends of the Earth (FOE) has built a solid reputation as a tough, innovative organization. Among FOE's cutting-edge projects are: a Washington, D.C. law banning the transportation of hazardous rail cargoes (like chlorine gas) through the city; a US national campaign to get dangerous chemicals out of cosmetics; and the exposure of fake press releases from the Department of the Interior. FOE's 'Green Scissors' program has helped federal, state and local governments save more than US$25 billion on projects that were polluting the air, land and water. In November 2008, FOE announced the completion of a merger with Bluewater Network. The merger of two of the most innovative and effective environmental organizations created a more powerful voice for the protection of the earth's peoples and the planet.

Bluewater Network has formulated a number of recommendations for action by the EPA to undertake as important first steps in a process to prevent further pollution of the marine environment by cruise ships. It has called on the EPA to:

- Conduct an in-depth assessment of the volumes and characteristics of the spectrum of waste streams from cruise ships.
- Analyze the potential impact of cruise ship waste streams on water quality, the marine environment and human health.
- Examine existing federal regulations governing cruise ship waste streams.
- Formulate recommendations on how to better control and regulate cruise ship waste streams.
- Initiate and actively pursue coordination and follow-up activities to ensure that any recommendations called for in this report which may not fall under the EPA's jurisdiction are implemented by the relevant agencies (e.g. the US Coast Guard).

Sustainable tourism: Tourism growth vs. tourism development

The growth of tourism in general, and in coastal areas in particular, is related to three main factors: increased personal incomes and leisure time, improvements in transportation systems, and greater public awareness of world destinations due to improved communications (EEA 2001). This growth, which has reached its peak in recent decades, exerts pressures on the environmental and cultural resources of coastal areas, and negatively affects the social, economic and cultural patterns of tourist destinations. Despite the prevailing growth trends, those who manage and invest in tourism are increasingly aware that it is sustainability in coastal areas that is strongly dependent upon the quality of these particularly fragile environments. Today's tourists seek a variety of experiences (including cultural and natural attractions, gastronomy and sports), all in a well-preserved and distinctive natural environment. At the same time, people living in traditional tourist destinations are increasingly aware of and concerned about their natural, historic and cultural heritage.

In order to satisfy both visitors and those who make a living from coastal tourism, the following are seen as priorities for a tourist development: limiting tourism growth, enhancing the tourism product, attracting a diversified clientele, and upgrading the quality of offerings and services. The natural resources and beauty of coastal areas have made them popular destinations for visitors. A healthy natural environment is a huge asset, but tourism generates pressures on local environment and ecosystems, such as higher water use, increased waste generation and accumulated emissions from air, road and sea transport in peak seasons. The Natura 2000 is an EU-wide network of nature protection areas, established under the 1992 Habitats Directive, aimed at ensuring the long-term survival of Europe's most valuable and threatened species and habitats, including in the marine environment.

The relationship existing between tourism and environment is best qualified as a relation of mutual dependence: not only is tourism highly dependent on environmental quality but environmental quality is also highly vulnerable to tourism development. Coastal communities certainly benefit from tourism through the creation of employment opportunities, the raising of revenue, the development of infrastructure, improvements in health and safety conditions, and enhancements of aesthetic standards. Raising the coastal populations' awareness of the value of their coasts under the influence of visitors and the subsequent investment in environmental protection are additional positive outcomes of tourism development.

Mass tourism is accused of being one of the main causes of ribbon development and severe ecological losses in coastal areas. Large tourism developments have dramatically altered not only the visual aspect of many coasts around the world but also the natural dynamics of coastal ecosystems. Land-grab and demand for resources are at the root of severe erosion phenomena, loss of valuable habitats (such as sand dunes, coral reefs, wetlands and mangrove forests), the irreversible destruction of pristine areas, and the loss of rare animal and vegetal species. Successful management of the sites allows their enjoyment as part of holiday experiences and their conservation for all future users.

Sustainable tourism development requires the informed participation of all relevant stakeholders, as well as strong political leadership to ensure wide participation and consensus building. Achieving sustainable tourism is a continuous process and it requires the constant monitoring of impacts, introducing the necessary preventive and/or corrective measures whenever necessary. Sustainable tourism should also:

> maintain a high level of tourist satisfaction and ensure a meaningful experience to the tourists, raising their awareness about sustainability issues and promoting sustainable tourism practices amongst them.
>
> (UNWTO 2004)

Innovative coastal tourism offerings

Many tourists nowadays seek a unique and customised experience rather than the more traditional type of 'sun-and-sea' package holiday. These changes in demand require reaction and adaptation by operators and destinations. The sector should develop new products promoting attractiveness and accessibility of coastal and marine archaeology, maritime heritage, underwater tourism, and enogastronomic experiences, among other innovative activities. The high quality of products and services is an essential source of competitive distinction.

The initiative Greek Challenge is following a new approach to promote sailing, by engaging in partnerships with businesses and universities to develop customised programmes. The initiative aims to bid for and host international sailing regattas, therefore contributing to the building of new, multifunctional infrastructure, extending the tourist period and introducing new know-how in the sailing sector. By using and enhancing maritime tourism knowledge, the initiative identifies the locations for regattas in an environmentally sustainable manner, examining what benefits such events would bring on a cultural, economic, infrastructure and social level. Islands and other remote destinations depend largely on boat services, which can represent a huge challenge for these destinations.

A plastic waste study

Researchers have warned that plastic waste could increase more than tenfold in the next decade. Whilst we know about the great Pacific garbage patch or 'gyre' and other huge swirls of plastic rubbish floating around in the world's oceans damaging marine ecosystems and sea creatures, it is not clear how much more plastic is making its way from land to sea every year. A recent study (Jambeck et al. 2015) calculated that about 8 million tons of plastic waste was in the world's oceans in 2010, and this is equal to five grocery bags filled with plastic for every foot of coastline in the world. The study indicated that plastic waste currently entering the ocean is one to three orders of magnitude greater than the reported mass of floating plastic debris in high-concentration ocean gyres and also globally. Furthermore, the researchers identified the major sources of this plastic debris and listed the 20 countries that are the biggest polluters. China tops the list, having sent an estimated 1.32 to 3.53 MMT of plastic debris into the sea; Indonesia comes a distant second at 0.48 to 1.29 MMT, followed by The Philippines, Vietnam and Sri Lanka. India is twelfth on the list, Brazil is at 16 and the US is number 20.

Combining data on solid waste from 192 different coastal countries with factors such as population density and economic status, Jambeck et al. (2015) found that the total annual waste generated by each country is mostly a function of its population size, with the top waste-producing countries having some of the largest coastal populations, which explains why so many Asian countries are on the list. Also, the largest contributors of plastic waste had the highest percentage of mismanaged waste. Mismanaged waste is a function of inadequate management, for example open dumping and litter. This mismanaged waste goes uncaptured, meaning that it then becomes available to enter marine environments.

Sixteen of the top 20 producers are middle-income countries, where fast economic growth is probably occurring but waste management infrastructure is lacking. Only two of the top 20 countries have mismanaged fractions <15%; here, even a relatively low mismanaged rate results in a large mass of mismanaged plastic waste because of large coastal populations and, especially in the US, high per capita waste generation. (Although US citizens create more plastic rubbish than their counterparts in middle-income countries, this trash is managed better.)

To stem the rising flow of plastic debris entering our ocean over the next decade, Jambeck et al. (2015) suggest that nations around the world need to reduce their overall waste and adopt better management strategies. We need to make sure that we are collecting and capturing solid waste and plastic around the world. Solutions will require a combination of local and global efforts. They need to be culturally appropriate and sensitive to social and economic concerns. However, a shift in how we manage waste could provide jobs and opportunities for economic innovation and it could improve the living conditions and health of millions of people. Responsible tourism and responsible tourists can contribute significantly to the reduction of waste and the protection of the environment.

Responsible tourism and responsible tourists

Travel and tourism should be planned and practiced as a means of individual and collective fulfillment. When practiced with an open mind, it is an irreplaceable factor for self-education, mutual tolerance, and learning about the legitimate differences between peoples, cultures and their diversity. Everyone has a role to play in creating responsible travel and tourism. Governments, business and communities must do all they can.

Guests can support responsible tourism in many ways to make a difference:

- Guests should open their minds to other cultures and traditions: it will transform their experience, and they will earn respect and be more readily welcomed by local people.
- Guests should be tolerant and respect diversity (i.e. by observing social and cultural traditions and practices).
- Guests should respect human rights. Exploitation in any form conflicts with the fundamental aims of tourism. The sexual exploitation of children is a crime punishable in the destination or in the offender's home country.
- Guests should help preserve natural environments. They should protect wildlife and habitats and should not purchase products made from endangered plants or animals.
- Guests should respect cultural resources. Activities should be conducted with respect for the artistic, archaeological and cultural heritage.

Travelling light: The paradox of sustainable tourism

A tourist flying economy class from Britain to Kenya and back generates around a tonne of carbon emissions, according to the International Civil Aviation Organization (*The Economist* 2015). No matter how many times he reuses his towels or sits on a composting toilet when he is there, he could never hope to offset the burning of all that jet fuel. So does that mean the very notion of 'sustainable tourism' is an oxymoron?

The phrase 'sustainable tourism' has three possible meanings. The first is ecological. Given the contribution that transport, especially by air, makes to global warming, on this definition it is almost guaranteed to fall short. The only truly sustainable holiday would be camping in the back garden eating berries. The second meaning is social. Ideally, when cultures meet and gain in mutual understanding the long-term benefits will be intangible, but real. The final meaning is economic. Tourists who step off the beaten track have a chance to help lift the poor out of poverty and encourage them to preserve their environments for financial gain. The question, therefore, is how much weight to give to each.

According to UNWTO, an estimated 1.1 billion international trips were made in 2014, a 4.4% increase on the year before. As popular destinations become overcrowded, more people seek places that remain comparatively unspoilt. Yet pristine wildernesses do not stay pristine for long once they are on the holiday trail. The paradox of sustainable tourism is that it can be both a destroyer of nature and an agent for its conservation. Keeping resorts small, and perhaps even temporary, can help resolve that paradox in favour of conservation.

Locals are too often persuaded to sell their land to developers for less than it is worth and villages can vanish once hotels are built. However, the Sandele resort in Gambia models another course: the resort is built on land leased from villagers, and it also employs them. When the lease runs out in 20 years' time the property will revert back to the locals, who now have the necessary management skills. The scheme also run courses for locals, including one on how to make 'rocket stoves' that require very little wood for fuel, thereby reducing deforestation.

Tourists on larger scale eco-tourism packages may be doing good work in other ways. Concentrating large numbers of visitors in a single location increases their local impact which can be for the better. If a resort buys local food, or invests in renewable-energy generation that can be used by those who live nearby, then the surrounding area can receive a boost. However, hotels must seek ways to mitigate their negative effects. Though signs suggesting that guests can help 'save the planet' by reusing their towels overstate the case, water-guzzling is one of the biggest evils of mass tourism. According to *The Economist* (2015), an analysis by Thomas Cook, a large holiday firm, suggests that on average each tourist around the world accounts for around 350 litres of water per day by showering and using the swimming pool, and this rises to 6,000 litres when indirect use such as food production is added. In Greece, for example, each tourist directly uses around three-fifths more water than a local.

Being more frugal with water can boost firms' profits. TUI, another big travel company, says it saved €2.2 million (US$2.5 million) in 2014 by cutting energy and water use at 43 of its hotels (*The Economist* 2015). Yet often it is the guests themselves who kick against energy-saving initiatives. To stop patrons leaving lights and air-conditioning on when they are out, many hotels have key cards that control the electrics in rooms. However, some guests override the system by inserting a business card into the control slot before exiting, rather than wait to recharge portable devices or tolerate a stuffy room for a few minutes on their return.

Overall, the benefits of sustainable tourism outweigh the harm. Often the most unnecessary flights are taken not by tourists but by business folk who fly abroad for a toe-touch meeting that could easily have been replaced by a video-call, and then fly home the same day.

CASE STUDY: EGYPT, THE NEW SUEZ CANAL PROJECT

Background

On the 5 August 2015, the newly elected president of Egypt, Abdel Fattah el-Sisi, inaugurated the building of an extra waterway alongside the existing Suez Canal. This very ambitious project is set to cost around EGP60 billion (US$8.4 billion), and is, according to the Suez Canal Authority, estimated to more than double revenues from US$5 billion to US$13 billion annually by 2023. Approximately 7.5% of world trade currently passes through the Suez Canal. The World Bank has announced its support for the project, and agreed to provide assistance and advice for the construction phase, which has already commenced. World Bank director Hartwig Schafer has stated that the project will change the economic landscape of Egypt and make foreign investment flow into the country.

The New Suez Canal Project is part of a major fiscal stimulus package meant to regain pre-2011 growth rates at 7% and to create jobs. It does not only involve the building of a second canal and improvement of the existing canal, but a Suez Canal Axis Development Project is also under way, which amongst other things involves the creation of an industrial hub in the adjacent area, the development of five new seaports, a technology valley, and a centre for supplies and logistics. This is meant to help diversify the economy towards more high-tech sectors, and divert demographic and economic concentration in Cairo towards the northeast.

A project of national pride

As with the Aswan Dam there has always, among the general populace of Egypt, been a sense of pride in this grand feat of engineering. It was therefore generally very well received, when President Abdel Fattah el-Sisi announced that the project will be funded exclusively by the Egyptian people through bank certificates, sold to Egyptian citizens residing both in Egypt and abroad. The popularity of the project has been apparent, as the entire amount of US$8.4 billion was attained in less than a week.

Growth and investment perspectives

The Egyptian economy remains in a fragile state, as the absence of tourism and investments continue to hollow out currency reserves and curtail growth. In order to prop up the economy, the Gulf countries of Saudi Arabia, The United Arab Emirates and Kuwait have provided Egypt with in excess of US$20 billion in aid and investments. Such funds are highly needed by an economy that is struggling with low domestic energy supply and a high budget deficit. These factors have so far kept international investors at bay, and the new government has therefore set forth plans to reform the extremely costly subsidy system, and to diversify the country's energy needs through wind, solar and nuclear power.

The government's goal is to bring the deficit down from 15% in January 2014 to 10% by June 2015. The positive tones in the Egyptian development have not passed international markets, and investors are steadily regaining confidence in the soundness of the Egyptian economy. So far 17.5 million cubic meters of waterway have been dug of the new canal, and 52 companies are present at the project site. The 72km long second canal involves 35km of dry digging and 37km of expansion and deepening. This will result in a reduction of the crossing time from 11 hours to 3 hours per ship, which accordingly will boost the daily capacity of the canal from 49 to 97 ships.

Consumer confidence stimulates trade

The demand for goods, both Egyptian and foreign, are increasing in Egypt, thereby contributing to the need for a new canal. Egypt was the first country in the world to build a man-made canal. The canal connected the Mediterranean Sea to the Red Sea in 1874 BC. As increased economic confidence is seen in Egypt, consumers in other markets are regaining confidence in their economies, thus spending is increasing and global trade is yet again on the rise, creating a favourable investment climate for the new Suez Canal Axis Development Project.

Some companies and countries have already expressed interest in investing in the new project. Maersk, with the majority share in the Suez Canal Container Terminal in Port Said, has expressed interest in further investments in Port Said as well as other potential projects as part of the Suez Canal Axis Development Project. Russia has likewise held talks with the government concerning investment in industrial zones.

Conclusion

The success of the New Suez Canal Project depends on the ability of the new government and ministries, in collaboration with international partners, to create a stable and attractive environment for investments, and to coordinate the project in a deliberate manner and within a realistic timeframe. If such criteria are met, the project could have an immense impact on the Egyptian economy and perhaps even on world trade.

Discussion

Sustainable has become a buzzword in the world of tourism. There is an overabundance of definitions but in this context, it is a business which attempts to make a low impact on the environment and local culture, while helping to generate income, employment, and the conservation of local ecosystems. It is also an ethos, an evolution in the way of doing things, and not a type of tourism product. It goes hand-in-hand (and is sometimes confused) with a number of similar types of tourism. It is important to establish the variations. *Responsible* tourism has similar principles to sustainable tourism, however it refers to the consumer's choice, i.e. the industry develops sustainable tourism and a responsible consumer chooses to participate in it. *Green* tourism is low impact, environmentally friendly tourism. While in general sustainable tourism is concerned equally with the social, economic and environmental impacts of tourism, green tourism focuses purely on environmental implications.

Versova beach, a one-and-a-half-mile strip of coastline in western Mumbai facing the Arabian Sea, has become known for the vast amounts of garbage littering its sandy shore. In October 2016, a small group of concerned citizens organized a clean-up, which spiralled into the largest beach clean-up effort in the world when it cleared more than £4 million of rubbish. However, while this ongoing initiative spearheaded by local residents is making steady progress at cleaning up this shoreline, we need to involve all levels of government and business in collaborative efforts to clear our planet of all waste, and to architect sustainable tourism.

References and further reading

Benoit, G. and Comeau, A. (eds) (2005). *A Sustainable Future for the Mediterranean: The Blue Plan's environment and development outlook*. London: Earthscan.

Bridges, T. (ed) (1997). *Travel Industry World Yearbook – The Big Picture 1996–97*. Volume 40. Spencertown, NY: Travel Industry Publishing Inc.

Business Research & Economic Advisors (BREA) (2015). *Economic Contribution of Cruise Tourism to the Destination Economies: A survey-based analysis of the impacts of passenger, crew and cruise line spending*. Vol. I (Aggregate Analysis) and Vol. II (Destination Reports). Prepared for Florida-Caribbean Cruise Association.

CoastLearn: www.coastlearn.org (accessed: April 2017). (A European internet-based vocational training package on Integrated Coastal Zone Management created in 2000 by EUCC in partnership with a large number of other organizations in Europe.)

The Economist (2015). 'Travelling light: The paradox of sustainable tourism'. www.economist.com/news/international/21662557-paradox-sustainable-tourism-travelling-light (accessed: April 2017).

Edvardsen, T. and OECD (2016). 'The future ocean economy'. Introductory Talk, MSEAS, Brest, 30 May 2016. www.ices.dk/news-and-events/symposia/MSEAS/Documents/MSEAS%20presentations/Edvardsen%20-%20TheFutureOceanEconomy.pdf (accessed: April 2017).

EEA (European Environment Agency) (2001). *Environmental Signals 2001*. Environmental Assessment Report No.8. Luxembourg: Publications Office of the European Union.

EEA (European Environment Agency) (2005). *The European Environment – State and outlook 2005*. Luxembourg: Publications Office of the European Union.

European Commission (2015). 'A European strategy for more growth and jobs in coastal and maritime tourism'. https://ec.europa.eu/maritimeaffairs/sites/maritimeaffairs/files/docs/body/coastal-and-maritime-tourism_en.pdf (accessed: April 2017).

Florida-Caribbean Cruise Association (2015). 'Cruise industry overview – 2015: State of the cruise industry'. www.f-cca.com/downloads/2015-Cruise-Industry-Overview-and-Statistics.pdf (accessed: April 2017).

GAO (US General Accounting Office) (2000). 'Progress made to reduce marine pollution by cruise ships, but important issues remain'. www.gao.gov/new.items/rc00048.pdf (accessed: May 2017).

Highlands and Islands Enterprise (HIE) (n.d.). *Awakening the Giant: A strategic framework for Scotland's marine tourism sector, 2010–2020*. Inverness: HIE.
Houston, J.R. (2008). 'The economic value of beaches – A 2008 update'. *Shore & Beach*, 76(3): 22.
INRA (EUROPE) (French National Institute for Agricultural Research) (1998). 'Facts and figures on the Europeans on holidays 1997–1998'. http://ec.europa.eu/commfrontoffice/publicopinion/archives/ebs/ebs_117_en.pdf (accessed: May 2017).
Jambeck, J.R., Geyer, R., Wilcox, C., Siegler, T.R., et al. (2015). 'Plastic waste inputs from land into the sea'. *Science*, 347(6223): 768–771.
Los Angeles Times (1999). 'Cruise line fined $18 million for dumping waste at sea.' http://articles.latimes.com/1999/jul/22/local/me-58520 (accessed: May 2017).
United Nations Environment Programme (UNEP) (2009). *Sustainable Coastal Tourism: An integrated planning and management approach*. Nairobi: UNEP.
United Nations Environment Programme (UNEP) (2013). *Green Economy in a Blue World*. Nairobi: UNEP.
United Nations University Institute for Water, Environment and Health (UNU-INWEH) (2012). *Land-Based Pollution Sources: A global analysis of land-based pollution sources science and transboundary management*. Hamilton, Ontario: UNU-INWEH.
United Nations World Tourism Organization (UNWTO) (2004). *Indicators of Sustainable Development for Tourism Destinations: A guidebook*. Madrid: World Tourism Organization.
United Nations World Tourism Organization (UNWTO) (2007) 'Tourism highlights'. www.e-unwto.org/doi/pdf/10.18111/9789284413539 (accessed: May 2017).
Wilson, P. and Wheeler, D.P. (1997). *California's Ocean Resources: An agenda for the future*. Sacramento, CA: The Resources Agency of California.

Chapter 13

Maritime clusters cooperating to compete

Definition of a cluster

> [A cluster is] a population of geographically concentrated and mutually related business units, associations and public (private) organizations centred around a distinctive economic specialization.
>
> (Porter 1990)

Firstly, a cluster is a *population*, not an entity. Consequently, the internal heterogeneity of clusters should be taken into account. Secondly, clusters are *geographically concentrated*. This dimension sets clusters apart from networks. Thirdly, the cluster population consists of business units, associations, public–private organizations and public organizations. Associations are defined as organizations that provide services to members. Associations are included in the population if they are cluster specific (i.e. if the majority of their members are included in the cluster population). Public and public–private organizations (such as education or research institutes) are included in the population if they engage in cluster-specific activities.

Fourthly, clusters are centred around a particular *economic specialization*, that can be regarded as the 'core' of the cluster. The notion of a core is implicit in most studies on clustering. Fifthly, clusters consist of business units and associations that are a part of, or relatively strongly related to, the core of the cluster. Such relations are in general both economic and social. Maritime clusters can contribute significantly to responsible ocean governance and the blue economy. However clusters do not happen naturally: they have to be constructed. Over the last decade, the concept of clustering has become a central concept for analyzing the competitiveness of nations, industries and firms.

Institutional collaboration

The Polish maritime economy, with about 12,000 enterprises, is an important part of the blue sector in the Baltic Sea Region (BSR). Gdynia Maritime University and the Polish Maritime Cluster at a BSR workshop presented ways to transfer knowledge between universities and enterprises. The most common activities include: internships and meetings of various scientific circles and individual students in enterprises with managers of companies; promoting preparation of engineering, master's and doctoral dissertations closely related to business practice; organizing seminars with the participation of representatives of the government, business and science; and conducting joint research projects. Gdynia Maritime University and the Polish Maritime Cluster took part in several international initiatives, established cooperation with many interested parties from the BSR, and represented the region on the annual meeting of the European Network of Maritime Clusters (ENMC).

The funding and establishment of the €90 million Namibe Fishery Academy in Angola is a good example of collaboration between the Angola Government, the Polish company Navimor International, Gdynia University and the Polish Maritime Cluster. The Academy is an institution of higher education with a maritime profile and educational specialists offering 4-year courses in three faculties. Navimor International and Gdynia Maritime University, with over 80 years of experience in maritime high-level education, have agreed to collaborate and deliver the educational and organizational concept of the Academy, educational and training programs for all three faculties, syllabi for all courses and specializations, specification of laboratory classes and lab equipment.

Maritime clusters unlock benefits of cooperation for the blue economy

A strong economy in Europe needs a strong and sustainable maritime sector and by forging connections between science, education and business, maritime clusters can help make this happen. The ENMC represents most of Europe's maritime industries via 17 national clusters. Clusters that are now all over Europe mean that businesses are able to: network on innovation, create economies of scale, and cooperate with each other nationally and internationally. The sum of their experiences can help the ENMC to move forward and feed into their decision making, helping them shape better policies. Maritime industries are at the heart of the EU's ambitions for blue growth, research, innovation and competitiveness. They account for about 5 million jobs and between 3.5% and 4.5% of the EU's GDP.

The ENMC welcomed the recent call for tenders regarding the study on the establishment of a framework for processing maritime economic data in Europe. The ENMC, together with the Directorate-General for Maritime Affairs and Fisheries (DG Mare), is one of the two driving forces behind this study since the organization stressed, repeatedly over the past years, the need for reliable European maritime economic data, as a paramount tool for helping to boost blue economy growth. Being dissatisfied with the inconsistency of existing data (including the questionable figures officially delivered by the EU on the occasion of the Limassol declaration in 2012), the ENMC made attempts to collect economic data from the 17 national maritime clusters of their network. However, the difficulties encountered were similar to the conclusions of the 2007 Eurostat study 'Approach towards an integrated maritime policy database' (see EC 2010), and did not achieve a reliable, homogeneous and satisfactory result. This situation then led ENMC to draft a guidelines document (Terms of Reference) which was sent to DG Mare in 2014. The Terms of Reference elaborated on the conditions for the success and usability of such a study, i.e. its independence, neutrality, the definition of common criteria, the possibility of updating the database on a periodic basis, and the inclusion of the ENMC and national maritime clusters as advisors in the peer review process.

A decision has recently been made by the European Commission to retain the principle of a pre-study, in order to identify and tackle the complications within a comprehensive data collection. This methodology of data collection is matching the expectations of the European maritime industry. Furthermore, the ENMC hopes that the maritime clusters will be well represented within the peer review board following the EC declaration in 2015: 'In the maritime field the advisory role of ENMC should be considered as necessary'.

US/EU cooperation arrangement

In 2015, a cooperation arrangement was signed between the United States Department of Commerce and the European Commission Directorate-General for Internal Market, Industry, Entrepreneurship and SMEs. The partners intend to collaborate in order to facilitate links between clusters in the US and the EU and to exchange best practices in fields of mutual interest in order to promote economic growth and job creation through increased trade, investment, research and innovation partnerships.

The participants intend to undertake the following activities:

- Promote EU–US cluster cooperation in the US and the EU at conferences, local events and meetings with local authorities, and cluster managers and cluster firms as an instrument for initiating contacts and identifying opportunities for cooperation between cluster partners.
- Include information on European cluster organisations and activities on the US Cluster Map and Registry website; and include information on US cluster organisations and activities on the European Cluster Collaboration Platform website.
- Link the European Cluster Collaboration Platform and the European Cluster Observatory websites to the US Cluster Map and Registry website; and link the US Cluster Map and Registry website to the European Cluster Collaboration Platform and the European Cluster Observatory websites.
- Regularly update relevant information about cluster developments in the EU and the US on the US Cluster Map and Registry and the European Cluster

Collaboration Platform websites, with specific sections publishing expressions of interest for cooperation.
- Encourage and facilitate the partnering between cluster organisations and cluster firms, in particular small- and medium-size enterprises, from the US and the EU, through the provision of information; and, in liaison with EU member states, facilitation of contacts and/or the organisation of international cluster missions (outgoing and incoming missions) offering matchmaking opportunities in strategic areas of mutual interest.
- Disseminate information on funding programmes that support international cluster cooperation activities for the benefit of clusters in the US and the EU.
- Exchange best practices on mutually beneficial topics on clusters of interest to both participants, such as cluster mapping, cluster collaboration, cluster internationalisation, and small- and medium-size enterprises' access to global markets through clusters, drawing upon best practices developed in other related EU–US forums and initiatives.

The cooperation anticipated under this arrangement is to be on a non-exclusive basis, and the arrangement may be modified or discontinued at any time by the participants in writing.

Collaboration in the Mediterranean

The Barcelona Convention entered into force in 1978, after Mediterranean countries and the EC had, 3 years earlier, adopted the Mediterranean Action Plan, the first ever Regional Seas Program under the umbrella of the United Nations Environment Program (UNEP). The convention was amended and renamed in 1995; and that version came into force in 2004. The 22 contracting parties to the Barcelona Convention were: Albania, Algeria, Bosnia and Herzegovina, Croatia, Cyprus, Egypt, the European Union, France, Greece, Israel, Italy, Lebanon, Libya, Malta, Monaco, Montenegro, Morocco, Slovenia, Spain, Syria, Tunisia and Turkey. In his opening speech at the Paris Declaration Convention for the Protection of the Marine Environment and Coastal Region of the Mediterranean in 2012, France's Ambassador for the Environment, Jean-Pierre Thébault, who chaired the meeting, said:

> In this very symbolic year for the environment, I express the wish that the Mediterranean Action Plan remains ambitious and leads by example, showing the way towards Rio+20.

At the UN Conference on Sustainable Development (Rio+20) in Brazil, the adoption of a strategic policy framework supporting a blue economy in order to safeguard and promote a clean and healthy environment was expected. The call came in a closing communiqué, the Paris Declaration, as the 17th Conference of the Contracting Parties (COP17) to the Barcelona Convention for the Protection of the Marine Environment and the Coastal Region of the Mediterranean and its protocols, drew to an end after 3 days of talks in the French capital on 10 February 2012.

Paris Declaration

The need for a 'blue' economy, a version of the green economy that is applied to seas and oceans, was underlined by the fact that the world's marine ecosystems provide essential food and livelihoods to millions of people. UNEP's research showed how a switch to the more sustainable green economy model could unlock the vast potential of the marine-based economy and at the same time significantly reduce ocean degradation while also alleviating poverty.

UNEP defines the green economy as one that results in improved human wellbeing and social equity, while significantly reducing environmental risks and ecological scarcities. The time has come to rethink how we manage our oceans, which are a key pillar for the economic and social development of many countries, and are vital in the fight against poverty. Yet too many of these essential natural resources are being degraded by unsustainable use, putting the ecosystems services they provide, such as food security and climate regulations, at risk. UNEP has further stated that management decisions and investments that prioritize the wellbeing of the oceans are essential if we are to continue to profit from this rich natural resource. The UNEP report 'Green economy in a blue world' (2012) looks at six different economic areas in the marine sector, and provides recommendations on how to boost their potential by implementing green measures.

UNEP's mission is to provide leadership and to encourage partnership in caring for the environment by inspiring, informing and enabling nations and peoples to improve their quality of life without compromising that of future generations. UNEP hosts the Coordinating Unit for the Mediterranean Action Plan of the Barcelona Convention. The Paris meeting welcomed the progress that had been made in 2011 in reinforcing the fight against deterioration of the Mediterranean sea with the entry into force of the Integrated Coastal Zone Management (ICZM) protocol and the protocol for the Protection of the Mediterranean Sea Against Pollution Resulting from Exploration and Exploitation of the Continental Shelf and the Seabed and its Subsoil (the Offshore Protocol). These protocols recognize the Mediterranean environment as a critical and shared resource, and promote a cooperative and holistic approach to its management. In the Paris Declaration, the contracting parties also agreed to:

- Reaffirm their political commitment to the sustainable development of the Mediterranean Sea and its coastal zones through an ecosystem approach to the management of human activities.
- Develop a coherent, well-managed network of marine protected areas in the Mediterranean, aiming for a target of 10% of marine protected areas by 2020.

- Intensify their efforts to curb marine pollution from land-based sources, such as mercury, persistent organic pollutants and marine litter, by adopting legally binding measures; and to reduce pollution from offshore and marine-based activities though regional action plans.
- Adopt the action plan for the implementation of the ICZM protocol, and encourage all contracting parties to ratify it.
- Protect the conservation and sustainable use of marine biodiversity in areas beyond national jurisdiction through the implementation of existing instruments and through the development of a multilateral agreement under the United Nations Convention on the Law of the Sea.
- Support the preparation by 2014 of a report on the state of the marine environment, including from a socio-economic perspective.

Danish Maritime Cluster benchmarking analysis

The Danish Maritime Cluster (DKMK) funded a benchmarking analysis in 2012 to measure its competencies against a range of internationally known and respected maritime clusters. By analysing the characteristics of these international clusters, the DKMK hoped to discover its own unique competence. Table 13.1 lists the eight clusters selected for the analysis, along with the five criteria used for measurement.

For this analysis, the maritime cluster is defined as shipping industry, transport-related services, maritime business services, and offshore, shipbuilding and equipment industry. Through the SWOT analysis, a strategy for further development of the competencies which already exist within the Danish maritime cluster would be outlined.

Although the analysis provided the possibility to uncover competencies lacking in DKMK, the emphasis has primarily been on developing the competencies that Denmark already holds, and often at a high level. Starting with a historic overview of the unique conditions related to the Danish maritime area, the analysis then further examined maritime policy that could have influenced the maritime development in a specific geographical area or country. The third benchmark in the analysis was the connection of sea to land, customers and end-users for the products that are transported by the maritime sector (i.e. infrastructure and logistics). The next benchmark was research and development already taking place in the cluster. Finally, the last benchmark was supply of higher education programs in and related to maritime activities.

The selection of the first two clusters for the benchmarking analysis, Norway and the Netherlands, falls on two geographically close clusters that are also old shipping nations, and that are very comparable to Denmark. Next is the cluster in London, chosen due to its status as a global financial centre and centre for maritime services, among these maritime insurance. Finally, the analysis looked outside of Europe, where Virginia on the East Coast of the US is one of the largest ports in the US, with a long tradition for trade and maritime activities.

Table 13.1 DKMK analysis of international clusters

Cluster	*Historic development*	*Maritime policy*	*Logistics*	*Research and development*	*Higher education*
Denmark	Strong and international traditions within shipping and shipbuilding companies. Recently, offshore activities have become a bigger part of the cluster.	Denmark's International Ship register/DIS; tonnage tax. Government action plans in 2006 and 2012.	Good infrastructure and connections to other countries. Ranked number 6 in LPI 2012.	Projects: the Danish Maritime Cluster and the Danish Maritime Fund. No dedicated national effort within the maritime area.	A specialized technical master program is under development. Technical and commercial electives. Executive MBA in shipping and logistics.
Norway	Shipping, shipbuilding, finance, classification, insurance and knowledge companies.	Norway's International Ship register/NIS; tonnage tax.	Inland transport dominated by shipping. Ranked number 22 in LPI 2012.	Publicly funded programs: Maritm21 and Smarttrans.	Bachelor and master programs in shipping, ship design, maritime law, logistics, etc.
The Netherlands	Historically a strong shipping nation. After World War II's destructions, focus is now on the strategic location and building of ports, infrastructure and logistics.	Manning Act; tonnage tax.	Rotterdam is Europe's largest port and logistics centre. Ranked number 5 in LPI 2012.	The industry-driven research centre: Dinalog.	Specialized masters degrees, especially at the Erasmus University in Rotterdam.
London, UK	Shipping and shipbuilding; but new leading knowledge-based cluster within finance, insurance, maritime law, etc.	British Ship Registry; request for support of education of seafarers.	London port: London Gateway. Ranked number 8 in LPI 2012.	UK Research Council; small dedicated public research funding.	Specialized master programs in numerous areas, especially maritime finance.

Virginia, US	Natural harbour; 400-year-old port on the US East Coast.	Public-owned harbour (Virginia Port Authority); outsourcing of terminal operations.	Access to large, densely populated areas in US dry ports; intermodal opportunities. Ranked number 9 in LPI 2012.	No access to information about specific public-funded projects other than what is done in the universities.	Special focus on education in natural sciences, for example engineering science.
Hong Kong	Natural harbour by the Pearl river. Focus from export–import to knowledge and hi-tech industry.	HK Ship Register; tonnage tax, favourable tax conditions and professional guidance; supply chain management centre.	'Green Lane' for trucks to Shenzhen. Ranked number 2 in LPI 2012.	PolyU research centre in shipping and logistics. Earmarked research funding.	A broad range of maritime programs at bachelor and master levels. Special focus on logistics and maritime studies.
Shanghai, China	Opened for international trade with the reforms in the 1970s.	'Privatization' of public-owned activities. Foreign shipping companies are almost equally treated to their Chinese competitors. Goal: international maritime centre in 2020.	The world's biggest container port. Ranked number 26 in LPI 2012.	Shanghai International Shipping Research Centre and Logistic Research Institute.	Shanghai Maritime University; special focus on shipping technology, IT and engineering science; 20,000 students.
Singapore	Independent since 1965 and has since then worked on making the unique location between East and West a strategic and economic advantage.	Good economic framework conditions. Maritime and Port Authority gives economic support to companies. Goal: global maritime knowledge hub in 2025.	Logistic centre with port–airport. Ranked number 1 in LPI 2012.	Public–private partnership with Singapore Manufacturing Federation and Maritime and Port Authority: Maritime Innovation & Technology (MINT) fund; Maritime Cluster Fund.	Master programs in maritime studies, maritime services, offshore technology, supply chain management, etc.

Source: Gammelgaard, B., Sornn-Friese, H., Hansen, J., Jessen, M., et al. (2013).

In Asia, the analysis looked at the two Chinese giants, Shanghai and Hong Kong. Furthermore, the analysis investigated Singapore, which has the same amount of inhabitants as Denmark and has developed its maritime cluster intensively during the last decades.

Data and information for this analysis was gathered by desk-research through the internet, reports, books and scientific articles. When explaining the position of the maritime clusters today, the different historical and geographical backgrounds are important considerations. Historical traditions dating back to colonial times is why London, the Netherlands, Virginia, Singapore and Hong Kong are significant maritime clusters today, even though they have evolved from pure centres for trade and shipping and maritime competencies to land-based service and knowledge-based maritime clusters.

Shanghai was open to international trade back in the imperial era, and the communist government thus considered it appropriate to invest in Shanghai as an international centre for trade and transport. Denmark's and Norway's maritime traditions are dated back to the Viking era. Today, both countries are offshore nations: Norway, of course, more than Denmark due to its geography and ownership of large oil fields in the North Sea. Denmark, however, is visibly different from the other clusters in the way that maritime activities take place outside of the country; in particular, after a large part of the shipbuilding industry has been relocated outside Denmark. Competencies in international trade and value-creation are thus particularly necessary for the further development of the Danish maritime cluster.

The maritime clusters in the analysis operate under very different political governance structures, which influence the degree of involvement of the public sector in the support of maritime cluster activities. The clusters span from Virginia in the US, where economic activities essentially are left to private initiatives (however, port terminals are public property), to Shanghai in China, where the government not only participates in planning and funding, but also strongly regulates the sector. Hong Kong and Singapore also have strong public strategic perspectives; however, private companies and their demands play a substantially stronger and clearer role. In London, public authorities participate in the establishment of new infrastructure (e.g. in the form of London Gateway), and the Dutch authorities are also active in the effort to establish a national, industry-driven research and development unit. Norway and Denmark both engage in the development of national strategies in order to improve global competitiveness.

Advantageous taxation systems, historically developed as a response to the growth in the flags of convenience, is a recurrent underling characteristic of the selected maritime clusters. In the maritime clusters of Virginia, London, the Netherlands, Singapore and Hong Kong, logistics and infrastructure play an important role; not least because these clusters are located around ports, where access and redistribution of freight are pivotal. The same applies to Shanghai, but logistical facilities and infrastructure to transport large volumes of freight do not play the same substantial role in Norway and Denmark (in Denmark due to the cluster's internationalization). Cooperation between companies, universities and public authorities is the driving force in the development of knowledge and innovation. The same development can be identified in Denmark, with regards to specific public funding of research and innovation.

A common feature of the maritime clusters in the analysis is the prevalence of higher education. In addition, general scientific disciplines can be applied in many companies within the maritime industry. The higher education concerning a specific maritime area is dominated by engineering; economy and management programs exist, but these are rarer. All in all, a future strategy for skill development in the Danish maritime cluster must be based on the strengths that the Danish maritime cluster already possesses. A considerable strength is the international outlook and commercial competencies; these competencies can sensibly be developed further as technical and economic cooperation skills in global supply chains and networks.

In Denmark, public authorities are active in the international policy context, and perhaps these strengths can be developed further. Moreover, based on democratic traditions, private Danish companies and public authorities are in a constructive dialogue, but this cooperation can be improved and competencies can be developed. Regardless of whether economic development will take place in Denmark (e.g. in the offshore activities in the North Sea) or abroad, the development of international competencies and technical and economic competencies on an international level are of absolute necessity. Today, clusters are tied to a geographic area, however, inside the maritime clusters companies act in global supply chains in global industries.

CASE STUDY: NORWAY, MARITIM21

Background

At the beginning of June 2010, a united maritime industry submitted Maritim21, an integrated strategy for maritime research and innovation, to Norway's Minister of Trade and Industry. The strategy was drawn up by the industry at the request of the Ministry of Trade and Industry between September 2009 and June 2010, in the course of a unique process in which several hundred people from both large and small companies in the sector and academic institutions participated in regional working meetings, work groups, direct interviews, and hearings via the internet. Maritim21's mandate was: to compile a holistic maritime research and innovation strategy for the twenty-first century; to develop new and more efficient forms of collaboration within the industry and between industry and politics; and to accelerate industry innovation with emphasis on new business models. As an integrated part of the government's Maritime Advisory Board (MARUT), Maritim21 is a result of over 5,000 hours of volunteer efforts by over 100 maritime businesses.

A Norwegian maritime research and innovation strategy for the twenty-first century

Norway is ranked as the world's fifth largest shipping nation, a leader in maritime operations for the petroleum industry, and has a maritime cluster which covers all commercial and technological aspects of the industry. This generates great potential for innovation. To maintain and strengthen Norway's position as a leading maritime nation in the next century, they must continue to develop effective, innovative and knowledge-intensive solutions. Maritm21 highlights some long-term challenges facing the industry and plots the course towards this goal.

Building on strengths

Maritim21's strategy concentrates on further enhancing areas in which Norway's maritime industry is already strong; high competence, smart solutions and competitive regulatory conditions. Seven priority areas were selected in the strategy process, divided into fundamental areas and application areas. Two of these areas had special priority in 2011: distribution and use of liquefied natural gas (LNG), and energy efficiency.

The seven priority areas

- *Knowledge hub and infrastructure:* Norway as a global maritime knowledge hub for the international maritime industry.
- *Maritime politics and regulations:* Knowledge is required to develop the most efficient and beneficial regulations.
- *Innovation and maritime business development:* Capture and develop the best ideas and bring them rapidly into the global markets.
- *Efficient energy:* Ensure that the Norwegian controlled fleet is the most environmentally friendly and cost efficient in the world.
- *LNG fuel:* Norway as a leading exporter of LNG technology and knowledge.
- *Demanding maritime operations:* Ensure Norway's position as prime operator of demanding maritime operations, both for design, equipment, ship/vessel and operations.
- *Arctic transport and operations:* Create values of arctic areas while ensuring safety of personnel and being considerate towards the environment.

As international shipping is facing stricter emission requirements, LNG will soon become a key marine fuel, due to its environmental and economic advantages. Norwegian shipowners, producers of maritime equipment, as well as maritime authorities have long been exploring the opportunities of LNG. Now they are aiming to increase the use of LNG both nationally and internationally. The main focus will be on encouraging more countries to facilitate the use of LNG fuel: reducing technological and logistical barriers, as well as optimizing and further developing the technological edge they have in this field. A broad and closely knit cooperation between all companies involved is the key to achieving Maritm21's strategic goals. In a gathering in Oslo of 85 participants from the industry, collaboration opportunities within the strategic areas of LNG and energy efficiency were discussed. The goal of the workshop was to coordinate applications to the Research Council of Norway and Innovation Norway, which will be important funding partners in the process. The companies involved agreed to cooperate on coordinated efforts where they can, whilst still competing fiercely when they have to.

Conclusion

The Maritim21 initiative has received a very positive response, both regarding its strategic direction and its model of cooperation. The strategy also calls for a considerable increase in financial input for maritime research and innovation projects, both from the government and from the industry itself. The industry is now clustering around the strategic priority areas to meet the challenges and to develop solutions jointly. Through targeted cooperation, they aim to create unique arenas for innovation and to reach the Maritim21 vision: Norway – the most attractive location for global, knowledge-based, environmentally robust maritime business.

Discussion

Clearly, the world has become more complex. Economic development, technology and globalization have all helped to blur the lines of old boundaries to such an extent that we desperately need to re-examine old rules, processes and practices. At the same time, machine intelligence is quickly replacing human cognitive power much like machines began to replace muscle power over a century ago. More and more, what drives value is the ability to collaborate with both humans and machines. The economy of today is therefore a social economy with collaboration at its centre. Domination by accumulating resources and driving efficiency is an outdated concept: today, we must focus on agility and interoperability. We need to shift our focus from assets and capabilities to empathy, design and networked organizations.

References and further reading

Blue Economy (2012). 'Countries call for blue economy to protect the Mediterranean'. www.blueeconomy.eu/blog/countries-call-for-blue-economy-to-protect-the-mediterranean/ (accessed: April 2017).

EC (European Commission) (2010). 'Approach towards an integrated maritime policy database'. https://webgate.ec.europa.eu/maritimeforum/en/node/1616 (accessed: May 2017).

EC (European Commission) (2012). *Progress of the EU's Integrated Maritime Policy*. Luxembourg: Publications Office of the European Union.

Gammelgaard, B., Sornn-Friese, H., Hansen, J., Jessen, M., et al. (2013). *Competencies in the Danish Maritime Cluster: A benchmarking-analysis*. Prepared by the Department of Operations Management and Centre for Shipping Economics and Innovation at the Copenhagen Business School. Copenhagen: Danish Maritime Cluster.

Johnson, R. (2012). 'Blue economy can protect Mediterranean Sea'. *Global Perspectives*, 13 Feb 2012. www.global-perspectives.info/index.php/international-edition/2012/march-incl-february/218-blue-economy-can-protect-mediterranean-sea (accessed: April 2017).

Porter, M.E. (1990). *The Competitive Advantage of Nations*. London: MacMillan.

Rodrigue, J-P. (2016). 'Strategic importance of maritime clusters in the ocean economy'. Presentation at the UNCTAD international seminar, Geneva, 10–12 May 2016. www.porteconomics.eu/2016/07/28/strategic-importance-of-maritime-clusters-in-the-ocean-economy/ (accessed: April 2017).

SmartComp (2012). 'Maritime cluster analysis on the Central Baltic region'. SmartComp Research Report No.1. www.utu.fi/fi/yksikot/tse/yksikot/PEI/tutkimus/Documents/SmartComp%20Research%20Report%201,%20December%202012.pdf (accessed: April 2017).

UNEP (United Nations Environment Program) (2012). 'Green economy in a blue world'. www.undp.org/content/undp/en/home/librarypage/environment-energy/water_governance/ocean_and_coastalareagovernance/a-green-economy-in-a-blue-world.html (accessed: April 2017).

Chapter 14

Constructing a local maritime cluster

The process of constructing a cluster

Constructing or building a cluster generally starts by selecting a cluster core. A cluster core should be relatively primary, in the sense that its location does not depend strongly on the presence of other economic activities. Furthermore, a core should consist of a spatial concentration of similar activities. Many firms in the same industry, a relatively large share of an industry in the regional product and a relatively high level of exports are indicators of spatial concentration. Secondly, the strength of linkages of the core with related organizations is analyzed. On the basis of this analysis, a list of cluster activities is developed. Thirdly, the relevant cluster region is delimited. Clusters can be local, regional or even inter-regional. Areas in the proximity of the core region where many economic activities are related to the cluster core are included in the cluster region. Fourthly, the cluster population is defined, consisting of business units associations and public (private) organizations that are both strongly linked to the core activity and located in the relevant cluster region.

The New Zealand model

This model follows an eight-step process for building a cluster:

1. *Analyse local economy:* Identify the embryonic and the more developed local clusters and prioritise the ones on which ones to initially concentrate.
2. *Initial cluster stock-take:* Gather information about the cluster in the local economy.
3. *Establish the leadership team:* Carefully choose the appropriate people to participate in the leadership group.
4. *Develop cluster vision:* Establish the preferred future for the cluster.
5. *Identify stepping stones:* Identify the key steps to the preferred future.
6. *Immediate action agenda:* Highlight the short-term projects.
7. *Institutionalize the cluster:* Set up an institution/organisation which will sustain the clustering process into the future.
8. *Upgrade the strategic agenda.*

Cluster building is already an accepted tool in economic development. As Professor Michael Porter has emphasised, 'Clusters are the building blocks of a productive, innovative economy' (Porter 1990). International experience indicates that the level of business formation tends to be higher in clusters. Start-ups are more reliant on external suppliers and partners, all of which they find in a cluster. Clusters therefore reduce the costs of failure, as entrepreneurs can fall back on local employment opportunities in the many other companies in the same field. Companies and research institutions can achieve higher levels of innovation, and closer interaction with customers and companies creates more new ideas.

The Dutch maritime cluster

The cluster concept is often used to analyze the economic specialization of a country. As previously stated, a cluster is a geographically concentrated set of firms related to a common product or process. This set of firms is homogenous: some firms play a more significant role in the development of the cluster than others. One specific set of firms, termed leader firms, are important drivers of the development of economic clusters.

Leader firms in the Dutch Maritime Cluster (DMC) were identified and the ways in which they contribute to the performance of the cluster were analyzed. Leader firms are a specific set of firms. Most leader firms are 'in between' ordinary small- and medium-sized firms and multinationals. Compared to multinationals, leader firms are much more embedded in their 'home nation'. The number of local partners in their innovation and production networks is relatively large.

The total added value produced by the maritime industry in the EU is €70 billion. The DMC generates 10% of the European value added. The share of maritime activities in the national product in the Netherlands is twice as high as the EU average (Policy Research Corporation 2008). This shows that the Netherlands is specialized in maritime activities. The DMC is extensively documented by studies commissioned by the Dutch Maritime Network. Studies by Peeters et al. (1999) identified 11 maritime sectors that make up the DMC.

The leader firm concept

The firms concentrated in a cluster are inter-related due to the existing buyer–supplier and knowledge-exchange relationships. Firms locate in clusters because the environment provides advantages: the proximity of customers and suppliers, a joint labour pool, and the presence of knowledge and information (Krugman 1991). Leader firms are firms with a relatively large impact both on other companies in the cluster and the cluster as a whole. Lorenzoni and Baden-Fuller (1995, p.147) define leader firms as 'strategic centres with superior co-ordination skills and the ability to steer change'. They distinguish four ways in which a leader firm contributes to the competitiveness of their partners: through strategic outsourcing, by sharing knowledge, by forming a bridge between different networks, and by focusing on competition on a value chain or network level rather than on a firm level. Lazerson and Lorenzoni (1999, p.362) identify 'focal firms', which are 'companies that occupy strategically central positions because of the greater number and intensity of relationships that they have with both customers and suppliers'.

The most prominent outcome of leader firms is the role they play in innovation. Leader firms can act as lead users (a concept discussed in von Hippel 1989); firms with an advanced demand that triggers innovation with suppliers. In the abovementioned studies on leader firms, their effects on the cluster as a whole are recognized, but the effects of leader firms on other firms in the cluster are not analyzed. Albino et al. (1999) stress the importance of a leader firm for the development of other firms in the cluster.

Leader firms can be enablers for the internationalization of other firms in the clusters. Albino et al. (1999, p.57) argue that 'leader firm internationalization can be considered the main impulse for district internationalization'. Leader firms act as 'launching customers' for the internationalization of their suppliers or sell products from the cluster in foreign markets. Other scholars have also recognized that the development of small firms depends to some extent on the presence and behaviour of a larger firm with strong coordination skills (see Kaufmann 1995; Lazerson and Lorenzoni 1999). Leader firm investments encourage innovation, enable internationalization of other firms in the cluster and improve the quality of the labour pool. In these ways, leader firms contribute to the competitiveness of other firms in the cluster and, as a consequence, the cluster as a whole.

Owing to their size, market position, knowledge and entrepreneurial skills, leader firms have the ability and incentive to make investments with positive externalities for other companies in the cluster. Positive externalities are central in the definition of leader firms. The concept of 'network externalities' is widely acknowledged (Economides 1996). The benefits of (additional) investments in networks (inter-firm, infrastructure and communication networks) amount to all network partners. Thus, in principle *all* investments that increase the competitive position of a network have positive externalities. Apart from these network externalities, cluster externalities (also termed agglomeration or localization-economies, see Richardson 1978; Krugman 1991) exist. Cluster externalities differ from network externalities because firms in the cluster benefit from these externalities, not just firms included in a relatively closed inter-firm network. The importance of externalities in clusters has been widely addressed (see Harrison 1992) and studied in the 'New Economic Geography School' (see Fujita and Mori 1997).

However, in these studies, the role of individual firms in creating these externalities is not widely addressed. Business scholars address the issue of firm networks (see Haakanson 1999; Boari and Lipparini 1999), since networks are relevant for understanding how firms (aim to) create competitive advantage. Yet these scholars do not focus on firm behaviour related to creating positive externalities. Investments of (leader) firms with substantial network externalities include investments in innovation and internationalization. The benefits of both innovation and internationalization spread to all members of the network. Three investments with substantial cluster externalities are identified: training and education provision, knowledge and information infrastructure, and an infrastructure for collective action. These investments improve the competitiveness of the cluster. Although the positive external effects can simply be side effects of investments of large firms that are not relevant for the investment decision of these large firms, creating externalities can also be a part of a strategy of large firms. In this case, leader firms can take the externalities into account in their decision processes.

The strong market position of leader firms can mean that they receive the largest part of the benefits of a more competitive network/cluster. In this way, leader firms have incentives to create positive external effects. Where internal benefits are sufficient for the leader firm to justify full investment, leader firms may aim to create arrangements to 'internalize the external effects', but their investment decision does not depend on it. Leader firms can also have an incentive to make investments with shared benefits, but not a sufficient incentive to invest by themselves. Such investment projects, provided that shared benefits exceed shared costs, face a 'collective action problem' (Olson 1971). For such investments, a financial arrangement to share costs is necessary. Given their position as prime beneficiary, leader firms have the incentive to play a leading role in devising such arrangements.

Maritime clusters and the quadruple helix system

There is a symbiotic relationship between maritime clusters and the quadruple helix system. The problem with the triple helix innovation model is that it is based on three pillars: academia, government and industry, which play integrated and sometimes overlapping roles. However, according to the quadruple helix innovation model, a country's economic structure lies on four helices: academia, industry, government and civil society; and economic growth is generated by the clustering and concentration of talented and productive people. A new setting for innovation has emerged: global networks of knowledge hubs, where sustainable innovation is based on ethically, socially, economically and environmentally sustainable principles. Networks contribute significantly to the innovative skills of firms by exposing them to new sources of ideas, by speeding up access to resources, and by facilitating the transfer of knowledge.

Clustering involves the interaction of businesses of a similar type, and the consequent fostering of healthy competition, networking and collaboration. Cluster theory emphasizes the role of networks and relationships between the various parts of a cluster but not of individual firms, and also fits with models of innovation and competitiveness. However, the benefits of clustering for the development of coastal and marine tourism, and for improving the competitiveness of the maritime sector, cannot be underestimated. At the level of ocean business, clusters help companies to take advantage of market opportunities that they could not achieve alone. Clusters can reduce the 'barriers' of cost and risk and adapt international rules and standards to the local context. Maritime clusters add value to the development of the economy by creating groups of networked businesses in specific sectors and improving business capabilities by offering tailored support. They enable economic specialization of a particular region, in a range of related activities.

This differentiated approach, targeted at a wide range of economic activities, enables clustering to take into account the variation in coastal and marine tourism structures and ocean business needs. Maritime clusters can contribute to an increase in economy-wide competitiveness by facilitating policy reform, fostering private–public dialogue and becoming a catalyst for wider private sector development initiatives. Recent decades have seen a shift from an earlier focus on innovation sources confined to a single institutional sphere (whether product development in industry, policy making in government or the creation and dissemination of knowledge in academia), to the interaction among these four spheres as the source of new innovative organisational designs and social interactions. Quadruple helix has grown into a conceptual framework for exploring the complex dynamics of the knowledge society and for informing ocean policymakers at national, regional and international levels in the design of new innovation and development strategies.

CASE STUDY: SINGAPORE MARITIME CLUSTER

Introduction

Maritime Singapore is a thriving ecosystem of maritime and port services, where the international maritime community congregates, and where ideas and opportunities abound. It comprises over 5,000 establishments employing more than 170,000 people, and contributes about 7% to Singapore's GDP (MPA 2009a). Singapore is a premier global hub port, connected to more than 600 ports in over 120 countries, giving users reliable, value-added and cost-efficient marine services and facilities. It is also an International Maritime Centre, home to more than 120 international shipping groups which continues to attract a diverse range of maritime businesses. Finally, Singapore is the Asian gateway for global leaders in ship financing, shipbroking, risk management and marine insurance.

Background

Despite the fact that Singapore is an island republic with limited natural resources and a small domestic market of approximately five million people, the country has experienced remarkable economic growth rates since its political independence in 1965, and is now counted as a highly developed and successful market economy. Singapore enjoys a GDP per capita of US$60,500 (2011 est.), which is equal to and higher than that of many developed Western nations (CIA n.d.). The maritime cluster is comprised of two key components: (1) core maritime sectors and (2) non-core maritime sectors. Core maritime sectors refer to entirely maritime-related activities such as shipping and port-related sectors, including offshore, shipbuilding and repair, shipping lines, etc. Non-core maritime sectors are those for which maritime-related activities form only a part of their total operations such as logistics and supporting services, engineering and technical services, and ancillary services (e.g. maritime finance, insurance, research and development, etc.).

The port as a trans-shipment hub

The Port of Singapore is one of the world's busiest ports in terms of trans-shipment and total containers handled. In 2011, total container throughput was 29.9 million TEU and total cargo tonnage was 531,175. Besides offering efficient cargo handling, Singapore is also one of the top bunkering ports (for ship refuelling), lifting 43,153 million tonnages of bunkers in 2011 despite the fact that Singapore does not produce oil (MPA 2011). Singapore is the world's third-largest petrochemical refiner and operates the most efficient and technically advanced ship-repair and shipbuilding facilities in Southeast Asia. Another factor contributing to the success of the Port of Singapore as a trans-shipment hub is its strategic location. In order for a port to be categorized as strategically located, it must have at least one of the following three characteristics:

- Situated on a main maritime route.
- Situated in or near production or consumption centres.
- Possessing natural deep water development possibilities.

Situated along the Straits of Malacca, the Port of Singapore is a main shipping route between East and West. Furthermore, Singapore is fortunate in possessing natural deep waters and ports, which allow it to service ships with deeper draughts without necessarily resorting to extensive and expensive dredging operations. Moreover, port operations are safe and reliable because Singapore enjoys favourable climatic conditions (i.e. no typhoons or other natural calamities) (Tongzon 2005). Located in one of the most dynamic regions in Asia, the port is close to some of the world's most dynamic economies, which has even further strengthened Singapore's position as a strategic centre for maritime business.

An International Maritime Centre

Built upon its status as a one of the world's busiest trans-shipment hubs, together with offering a range of maritime services, Singapore is designated an International Maritime Centre (IMC). It is the mission of the Maritime Port Authority (MPA) to attract shipowners and operators, and to encourage them to establish their operations and regional headquarters in Singapore, based on the government's pro-foreign direct investment policies, a business-friendly tax system, excellent infrastructure and the supply of maritime services. What is more, supported by maritime funds and in close collaboration between academic institutions and the maritime industry, Singapore strives to further develop expertise, skills and know-how through investing in maritime research and development (R&D), and to become a global maritime knowledge hub by 2025. The MPA's mission was established in 1996: to advance and safeguard Singapore's strategic maritime interests, and to develop Singapore as a premier global hub port and IMC.

According to the World Bank, Singapore leads a list on the overall ease of doing business, followed by Hong Kong (World Bank 2012). In addition, according to the 'World competitiveness report 2012–2013', Singapore retains its second place as a result of an outstanding performance across the entire index. In particular, Singapore's public and private institutions rank as the best in the world for the fifth year in a row. The report also highlights Singapore's world-class infrastructure. Moreover, its strong focus on education has translated into a steady improvement in the higher education and training pillar, reinforcing Singapore's competitiveness (World Economic Forum 2012).

Maritime policy

The successful development of the Port of Singapore as a trans-shipment hub and the maritime cluster as a whole should be attributed not only to the strategic geographical location but also to the role of the Singapore government. Proactive government intervention and effective implementation of selective strategies and polices has created an environment of openness, efficiency and accountability in its port operations and services. Singapore's style of government intervention is very characteristic: while it encourages competition, efficiency and openness, it also intervenes in the economy both directly through government agencies and indirectly by way of regulation.

For example, the Port of Singapore is self-financing, runs on a commercial basis and is expected to compete with other ports on an equal basis. However, the port is also a public port and it is therefore expected to operate with objectives that are consistent with the national development agenda and priorities of the government (Tongzon 2005). By focusing on and supporting the diverse business needs of enterprises within maritime sector as well as related sectors, the government has created a pro-business framework. Furthermore, the government aims to constantly respond to the maritime sector by reviewing the tax system, incentives and other regulations. In fact, with six bilateral shipping agreements and seven free trade agreements alongside a favourable tax regime with more than 50 double taxation agreements, Singapore offers a hugely business- and tax-friendly environment (MPA 2009b).

Ship registration

The Singapore Registry of Ships (SRS), established in 1966 and administered by MPA, is internationally recognized as a quality registry of shipping, including efficient services and high operational and pollution control standards. In addition, SRS has good safety records. With no restriction on trading areas of Singapore-flagged ships together with a business-friendly environment, shipowners around the world enjoy both tax advantages and financial incentives to register their vessels under the Singapore flag. Subsequently, with more than 3,000 registered vessels and a total of 43.7 million gross tons, SRS ranks in the top ten of the largest registries in the world (MPA 2009c).

Infrastructure

Infrastructure is highly developed in Singapore, and this enables efficient handling of goods from vessels. Tongzon (2005) divides Singapore infrastructure into two categories: (1) physical infrastructure and (2) soft elements in infrastructure. The physical infrastructure consists of roads, railways, cranes and storage capabilities. Singapore is physically connected to Malaysia via railway and also by road to the rest of Southeast Asia with maritime links to destinations around the world. In addition, Singapore also has an important international airport, Changi Airport, which connects Singapore with 190 destinations worldwide and handles 70 million passengers yearly (Gammelgaard et al. 2013). The soft infrastructure (i.e. all the services required to maintain economic, social and cultural standards) is highly important in order to secure Singapore's 24-hour service. Singapore is a knowledge-based cluster, consisting of tangible and intangible knowledge competencies. Universities and R&D facilities are important institutions in the knowledge infrastructure (Kuchiki and Tsuji 2010).

In recent times, the Port of Singapore has faced significant competition from ports in China and Malaysia, where there are lower land and labour costs. Singapore has defended its position by investments in automation and ICT. The Port of Singapore is now one of the most automated ports in the world in terms of administration and organizing. An electronic data interchange system was developed, making administration and organization of berths in the port automatic. Furthermore, the CITOS system (Computer Integrated Terminal Operations System) automatically handles operations (Tongzon 2005).

Considered as a logistics centre, the Port of Singapore is a collective of two commercial terminal operators: The PSA Corporation Limited and Jurong Port. The PSA is categorized as container terminals and consists of Tanjong Pagar, Keppel, Brani, and Pasir Panjang terminals, with Pasir being the newest and capable to handle +13,000 TEU. Jurong Port was established in 1965 and is a multipurpose port, handling all types of cargo, including steel, forestry products, metals, rubber, machinery and mechanical appliances, and project cargo. The multiple functions of Jurong Port enable it to handle various types of goods and to re-export goods or distribute it within the region. The throughput per annum was 570,573 TEU in 2011 (Jurong Port n.d.).

Research and development

Singapore's position as a leading international maritime centre is mainly built upon the pro-business framework. However, investing in the latest port and maritime technology together with supporting education, training and upgrading of workforce skills is also a core element in Singapore's strategy to ensure and reinforce competitiveness. The driving force behind Singapore's port and maritime development is the MPA, taking on the roles of Port Authority, Port Regulator, Port Planner and National Maritime Representative.

Industry and other agencies partner the MPA in order to facilitate port operations and growth, to expand the cluster maritime ancillary services, and to promote R&D and manpower development. In 2009, Singapore announced its vision to become a global maritime knowledge hub by 2025 characterized by excellent maritime R&D, resulting in innovative initiatives that the maritime sector can leverage both technically and commercially. In this regard, technology companies engaged in innovative maritime R&D and interested in placing operations in Singapore can receive financial aid during the early stages of product development. Moreover, MPA will support facilitating and commercializing the inventions. In its commitment to develop Singapore into a dynamic centre of excellence for maritime R&D and to strengthen the maritime cluster, MPA has also established partnerships and collaborations between tertiary and research institutions and the maritime industry. This framework involves the following key areas:

- Education and training to develop R&D expertise.
- R&D to generate intellectual property and results.
- Test-bedding for the prototypes from R&D.
- Commercialization of the developed products and services.

To support the implementation of the abovementioned areas, MPA has established an S$100 million Maritime Innovation & Technology (MINT) fund to encourage local and foreign companies using Singapore's port and maritime facilities to undertake maritime R&D. A central element of MPA's strategy to sustain Singapore's growing maritime cluster is the development of a continuous supply of highly skilled manpower. Hence, in 2002 MPA introduced the Maritime Cluster Fund (MCF), which supports and sponsors a range of programs to develop maritime expertise. In practice, to encourage maritime companies to train and develop the capabilities of their employees, the S$50 million MCF solely focuses on manpower development, covering the whole of the maritime cluster, including ship owning/operating, ship management, ship financing, ship brokering/chartering, maritime law/arbitration, marine insurance, port terminal management planning and development, and marine engineering and naval architecture.

Another important actor within the maritime cluster is the Singapore Maritime Foundation (SMF: www.smf.com.sg), which was established in 2004 with the purpose of forging partnerships between the public and private sector. Functioning as the link between the government and the private maritime sector, it is the objective of SMF to promote Singapore as an IMC and to develop manpower to support the maritime cluster. The presence of these multiple agencies and stakeholders ensure an integrated development approach in Singapore.

Higher education

Maritime education in Singapore includes a wide range of higher education programs and continuing training courses from shipping-based courses, such as maritime economics, to technical courses, such as naval architecture. What is more, Singapore is home to a whole range of highly skilled and specialized workforces in maritime professional services such as shipbroking, shipping finance and maritime law.

For shipboard personnel, including officers and crew, the Integrated Simulation Centre (ISC) offers advanced simulators to provide efficient and realistic training. The ISC is also capable of conducting marine R&D for port planning. Finally, the Singapore Maritime Academy (SMA) offers part-time and diploma programs within Shipping Operations and Management and in Supply Chain Management.

Conclusion and discussion

Successful maritime cluster development entails a combination of parallel and complementing management processes. Using ecosystems to coordinate maritime stakeholders' efforts towards meeting specific challenges is becoming more prevalent in the twenty-first century economy. Understanding and maximizing effective use of ecosystems is the strategy that maritime actors may also adopt in order to operate sustainably. Coordination of the emerging maritime ecosystem activities, stakeholder group coordination, and change management is thus considered. Refined 'orchestration' efforts should be emphasized that create the necessary synergy between different sector activities, which are traditionally carried out in separate projects and incompatible working cultures. This process includes the clear identification of: the players and their business models; the goals that unites all the players; and the ecosystem roles, such as resources owners, problem solvers and facilitators, and process orchestrators. The rules of the ecosystem, the formal and informal relationships, and the challenges and opportunities that render the ecosystem or cluster approach appropriate should also be identified.

References and further reading

Albino, V., Garavelli, A. and Schiuma, G. (1999). 'Knowledge transfer and inter-firm relations in industrial districts: The role of the leader firm'. *Technovation*, 19: 53–63.

Boari, C. and Lipparini, A. (1999). 'Networks within industrial districts: Organising knowledge creation and transfer by means of moderate hierarchies'. *Journal of Management and Governance*, 3: 339–360.

CIA (Central Intelligence Agency) (n.d.). 'The world factbook'. https://www.cia.gov/library/publications/resources/the-world-factbook/index.html (accessed: April 2017). The Danish Shipowners' Association (2010). 'The economic significance of maritime clusters: Lessons learned from European empirical research'. http://ec.europa.eu/competition/consultations/2012_maritime_transport/dsa_3_en.pdf (accessed: April 2017).

Doloreux, D. and Shearmur, R. (2009). 'Maritime clusters in diverse regional contexts: The case of Canada'. *Marine Policy*, 33: 520–527.

Economides, N. (1996). 'The economics of networks'. *International Journal of Industrial Organization*, 14(6): 673–699.

Fujita, M. and Mori, T. (1997). 'Structural stability and evolution of urban systems.' *Regional Science and Urban Economics*, 27: 399–442.

Gammelgaard, B., Sornn-Friese, H., Hansen, J., Jessen, M., et al. (2013). *Competencies in the Danish Maritime Cluster: A benchmarking-analysis*. Prepared by the Department of Operations Management and Centre for Shipping Economics and Innovation at the Copenhagen Business School. Copenhagen: Danish Maritime Cluster.

Gao, M. (2014). 'Maritime clusters in China'. Electronic Publications of Pan-European Institute. www.utu.fi/en/units/tse/units/PEI/reports/Documents/Minghui%20Gao%20 3_2014.pdf (accessed: April 2017).

Haakanson, H. (1999). 'Learning in networks'. *Industrial Marketing Management*, 28: 443–452.

Harrison, B. (1992). 'Industrial districts: Old wine in new bottles?' *Regional Studies*, 26: 469–483.

Hippel, E. von (1989). 'New product ideas from lead users'. *Research Technology Management*, May–June.

Jurong Port (n.d.). Website. www.jp.com.sg/ (accessed: April 2017).

Kaufmann, F. (1995). 'Internationalisation via co-operation strategies of SME'. *International Small Business Journal*, 13(2): 27–32.

Kleywegt, A. and Goh, M.L. (2002). 'Competition between the ports of Singapore and Malaysia'. www.researchgate.net/profile/Guangyan_Wu/publication/265114505_Competition_between_the_Ports_of_Singapore_and_Malaysia/links/547ea4f60cf2c1e3d2dc20f2/Competition-between-the-Ports-of-Singapore-and-Malaysia.pdf (accessed: April 2017).

Krugman, P. (1991). *Geography and Trade*. Leuven, Belgium: Leuven University Press.

Kuchiki, A. and Tsuji, M. (2010). *From Agglomeration to Innovation: Upgrading industrial clusters in emerging economies*. Basingstoke, UK: Palgrave Macmillan, p.50–117.

Langen, P.W. de (2002). 'Clustering and performance: The case of maritime clustering in the Netherlands'. *Maritime Policy and Management*, 29(3): 209–221.

Lazerson, M. and Lorenzoni, G. (1999). 'Resisting organizational inertia: The evolution of industrial districts'. *Journal of Management and Governance*, 3: 361–377.

Lorenzoni, G. and Baden-Fuller, C. (1995). 'Creating a strategic centre to manage a web of partners'. *California Management Review*, 37(3): 146–63.

McKinnon, A. (2011). *Hong Kong and Singapore Ports: Challenges, opportunities and global competitiveness*. Hong Kong: School of Law, City University of Hong Kong.

Monteiro, P., de Noronha, T. and Neto, P. (2013). 'A differentiation framework for maritime clusters: Comparisons across Europe'. www.mdpi.com/2071-1050/5/9/4076/pdf (accessed: April 2017).

MPA (Maritime Port Authority) (2009a). 'Singapore to be a global maritime knowledge hub by 2025'. Press release. www.mpa.gov.sg/web/portal/home/media-centre/news-releases/detail/e3608fad-8821-419b-a0c5-d14e07ca14af (accessed: May 2017).

MPA (Maritime Port Authority) (2009b). 'Conducive business environment'. www.mpa.gov.sg/web/portal/home/maritime-singapore/what-maritime-singapore-offers/pro-business-environment/conducive-business-environment (accessed: May 2017).

MPA (Maritime Port Authority) (2009c). 'Singapore registry of ships'. www.mpa.gov.sg/web/portal/home/singapore-registry-of-ships (accessed: May 2017).

MPA (Maritime Port Authority) (2011). 'Annual report 2011'. www.mpa.gov.sg/web/wcm/connect/www/ac1bc1bf-ac83-4d3a-a6d8-a20802a33a9e/mpa_2011-annual-report-with-financials.pdf?MOD=AJPERES&CACHEID=ac1bc1bf-ac83-4d3a-a6d8-a20802a33a9e (accessed: May 2017).

Myles, P.B. (2017). 'The symbiotic relationship between maritime clusters and the Quadruple Helix System'. Proceedings of the 9th International Coastal & Marine Tourism Congress, Gothenburg, West Sweden, 13–16 June 2017.

Olson, M. (1971). *The Logic of Collective Action: Public goods and the theory of groups*. Cambridge, MA: Harvard University Press.

Peeters, C., Lefever, A., Linden, J. van der, Bruynseels, S., et al. (1999). *De Nederlandse Maritieme Cluster: Economische betekenis en structuur*. Delft, Netherlands: Delft University Press.

Policy Research Corporation (2008). 'The role of maritime clusters to enhance the strength and development of European maritime sectors'. Study commissioned by the European Commission (DG MARE). https://ec.europa.eu/maritimeaffairs/sites/maritimeaffairs/files/docs/body/report_results_en.pdf (accessed: May 2017).

Porter, M.E. (1990). *The Competitive Advantage of Nations*. London: Macmillan.

Reve, T. (2009). 'Norway – A global maritime knowledge hub'. http://web.bi.no/forskning/papers.nsf/b63c43a0385ff61dc1256e620043c6b0/ff9bf9873e2f5db2c125767a0034b521/$FILE/2009-05-reve.pdf (accessed: April 2017).

Richardson, H.W. (1978). *Regional and Urban Economics*. Oak Brook, IL: Dryden Press.

Sheffi, Y. (2012). *Logistics Clusters, Delivering Value and Driving Growth*. Cambridge, MA: MIT Press.

SinHoon, H. (2012). 'Building a logistics/supply chain hub – The Singapore experience'. www.elsalvadorcompite.gob.sv/portal/page/portal/ESV/Pg_Biblio_logist/MFA%20-%20Building%20a%20Logistics%20Supply%20Chain%20Hub.pdf (accessed: April 2017).

Sornn-Friese, H. and Gammelgaard, B. (2012). 'Maritime economics and management studies at the business school'. *Mercator: Maritime Innovation, Research and Education*, December: 488–495.

Tenold, S. (2010). 'A most convenient flag – The basis for the expansion of the Singapore fleet, 1969–82'. *Maritime Policy & Management*, 30(3): 255–268.

Tongzon, J. (2005). 'Key success factors for transshipment hubs: The case of the port of Singapore' in: K. Cullinane and T.-W. Lee (eds), *World Shipping and Port Development*. Basingstoke, UK: Palgrave Macmillan, p.163–175.

Tongzon J. and Heng, W. (2005). 'Port privatization, efficiency and competitiveness: Some empirical evidence from container ports (terminals)'. *Transportation Research: Part A*, 39(5): 405–424.

World Bank (2012). 'Doing business 2012: Doing business in a more transparent world'. www.doingbusiness.org/reports/global-reports/doing-business-2012 (accessed: May 2017).

World Economic Forum (2012). 'World competitiveness report 2012–2013.' http://reports.weforum.org/global-competitiveness-report-2012-2013/#cover (accessed: May 2017).

Chapter 15

European maritime cluster networks

EU policy

Europe's dynamic maritime sector is a world player. In shipping and logistics, shipbuilding and marine equipment, shipbroking and insurance, offshore energy, dredging and extraction, hydraulic engineering, fisheries, coastal tourism, and marine research, European expertise and European assets generate jobs at home and exports abroad.

Clusters promote the fullest development of the potential of the maritime sector, with their emphasis on cross-industry linkages and complementarities. Special skills acquired in offshore technology, for instance, feed into innovations in specialized vessels for deep-sea exploration, into the exploitation of the new world of 'blue' marine-linked biotechnology, or into investigation of methods for safe extraction and exploitation of energy from methane hydrates.

Effective collaboration in clusters can help to respond to environmental considerations: for instance, by applying high-tech engineering and design to alternative ship propulsions or ballast water-free ships. Combining complementary activities by marine research institutes can also boost understanding of the impact of maritime activities on the environment and how to minimize damage.

Some clusters focus on single geographic regions, often around a major port, while others are multicentred, bringing together expertise and experience from several coastal sites, often across national boundaries. Clusters can act as focal points for maritime expertise, to ensure continuity in skills and the preservation of knowledge. They can improve links between training and maritime industries, and promote the profile of the sector so as to help attract candidates into nautical colleges and maritime professions.

Today's players are subject to world competition. To stay ahead, Europe's diverse maritime activities need to maximize the possibilities for reinforcing one another, particularly since many of the firms involved are relatively small. Effective European policies can be beneficial in the following ways:

- A European policy can help by fostering the creation of maritime clusters, in line with the objectives of the EU's Lisbon Strategy.

- Maritime potential and regional development can be driven forward with the appropriate policy framework.
- Generating joint initiatives among different actors from the maritime economy, the research community and public authorities can improve access to resources and competences.
- A comprehensive maritime policy can ensure the best interaction between all related EU policies such as transport, industrial research and training. Regional and environmental policies all have their part to play in promoting the reinforcement that clusters can provide in the maritime sector.

EC study of the role of clusters

The maritime industries throughout Europe contribute to the wellbeing of all Europeans as the volume of transportation is determined by the world trade of goods. The majority of external trade of the EU is transported by sea. European ports handle almost 3.5 billion tonnes of cargo per year and more than 350 million people pass through these ports. Consequently, maritime transport (shipping) and its related services fulfill a vital role in the European landscape. The maritime works sector provides dredging services and land reclamation. For energy, 40% of oil and 60% of gas consumed in Europe is drilled offshore. The seas around Europe also provide a range of energy transport routes, via shipping, submarine pipeline networks, and electricity interconnectors. They are also important for carbon-neutral energy generation, through the rapid development of offshore renewable sources of energy, and the seabed's potential for permanent CO2-storage. Europe's coastal regions are home to maritime industrial activities, such as shipbuilding, and they are also among the world's top destination for tourists. With an annual average growth rate of 10%, cruise tourism is a booming industry. The fisheries sector provides a major food source and is an important employer in certain regions of Europe.

Europe therefore has an important maritime industry with a strong global position in many (sub-) sectors. The question is how to maintain and strengthen the competitiveness of the maritime clusters even more.

Over the last decade, in most member states, there has been an increased attention towards the maritime sectors and, meanwhile, many countries moved in the direction of having their maritime industries represented by a cluster organization. In November 2005, a European Network of Maritime Clusters (ENMC) was established: the national cluster organizations meet regularly and coordinated efforts to promote the European maritime sectors are worked out. The European Commission (EC) has recognized the importance of maritime cluster initiatives with regard to competitiveness and innovation, and the Blue Paper.

The EC (2008a) conducted a study of the role of maritime cluster organizations, as the entities to enhance cluster awareness. An overview of the European maritime cluster organizations, their success factors and their role in developing cross-cutting approaches and tools in the maritime sphere was considered necessary.

With this information, in addition to a clear mapping of the maritime clusters within the framework of the European Cluster Observatory, it would be easier to work out relevant future policy initiatives.

Main objectives of the study

The main objective of this EC study was to enhance the exchange of knowledge and experience: knowledge on the size, specialization and focus of the maritime sectors or clusters (mapping), and experience within the different cluster organizations. Ultimately, this would provide building blocks for evidence-based policy development in line with the Blue Paper and its aim to contribute to sustainable development and the competitiveness of the maritime sectors.

The report presented the following key results:

- An overview of European maritime clusters showing their main economic features.
- An assessment of success factors and the future role of maritime clusters in maritime policy development.

Economic inventory of the European maritime clusters

A cluster is defined, by Professor Michael Porter, as a geographically proximate group of interconnected companies and associated institutions in a particular field, linked by commonalities and complementarities (external economies). External economies that occur within a cluster are the economic and financial inter-sector relations, a common knowledge and technology base, and a shared labour market. The European maritime clusters have a strong position in the maritime world. The following examples substantiate this claim:

- European ports handle almost 25% of world seaborne trade.
- European shipowners control almost 40% of the world fleet.
- Europe is the world's number one tourist destination, with coastal tourism being one of its main attraction pools.
- European shipbuilding has been the region with the highest global turnover for most of the last decade.
- European yacht builders produce 60% of the mega yachts.
- European dredging companies have 80% market share of the open market.
- Europeans dominate the market for renewable offshore energy.
- European services, maritime research, inland shipping, fisheries and navy are world-leading sectors.

Regarding the economic inventory, with detailed focus on direct added value and employment, maritime sectors are divided into clusters in order to focus on developing a European cross-cutting policy approach for the sea-related sectors (combining offshore and coastal activities) such as: traditional maritime sectors; coastal (and marine) tourism and recreation; and fisheries. Differences between the employment figures used in the study and the employment figures originating from national maritime cluster organizations were analyzed. The role of a maritime cluster surpasses its own economic impact; in addition, it plays a key role in facilitating the functioning of the entire economy, for example by means of maritime transport facilitating international trade, and in generating indirect effects through purchases in the value chain.

In a study by the Policy Research Corporation (2001), which focused on the economic analysis of the traditional maritime sectors in member states and Norway, it was found that (after correction for double counting and cross-border effects) for every euro direct added value, €0.59 indirect added value was realized. Translating this ratio to the €186.8 billion added value in all sea-related areas in the current study would result in an indirect value of about €110 billion. This figure is only a rough estimate of the actual indirect economic effects (i.e. it gives a first idea on the size of these effects), and should be treated as such. The total economic impact of maritime activities may be broken down into direct and indirect effects. Direct effect considers the economic activity and jobs generated directly by the cluster. Indirect effects include the jobs and demand created up and down the supply chain, as well as the consumption generated in the rest of the economy. Thus, the value and demand generated in the clusters trickles down through the overall economy to create further jobs and demand, making the combined economic importance much greater than simply direct activities.

Specific maritime sector trends in Europe

- *Recreational boating:* This industry has enjoyed years of steady growth. In the coming years, a further annual growth of 5 to 6% is estimated. Reasons for this growth are the baby boom generation which is reaching its retirement age, increase of wealth, and the potential in emerging economies.
- *Offshore supply:* The EU offshore supply industry is closely related to the offshore exploration and production of oil and gas. European offshore oil and gas are mainly extracted in Norway, the UK, France, the Netherlands and Denmark. The depletion of these resources in the course of the next decades, as well as price fluctuations, will have an impact on the exploration and production activities and thus on the offshore supply industry. In that respect, a shift towards supplying alternative energy producers (e.g. wind turbine parks) is likely to take place.
- *Coastal tourism:* Although in particular Mediterranean regions coastal tourism still plays a major role, the European coastal tourism on the whole is facing increased worldwide competition. Trends in coastal tourism are: the demand for higher quality at the lowest possible price; the combination of sea-related activities with the natural and cultural heritage of the hinterland in the Atlantic and Baltic regions; the increasing concern for the preservation

of cultural heritage; and the increasing cost of maintaining basic natural resources for tourism such as beaches and other coastal amenities.

- *Cruise shipping:* The world cruise market continues to grow and the market share of European cruise passengers is rising (it currently amounts to an 11% share of the global market). The European cruise market has become the world's second largest destination after the Caribbean. The cruise sector is a European growth industry: between 1995 and 2005 European cruise tourism increased by 230% to 3.3 million passengers. The cruise line industry is becoming increasingly concentrated due to mergers and acquisitions.

European maritime policy initiatives

On 10 October 2007, the EC launched its Integrated Maritime Policy (IMP), setting out its vision for a sustainable future for the oceans and seas in the 'Blue Paper'. Among its aims are to raise the competiveness of Europe's maritime economic activities and to make the most of Europe's potential for research and innovation. It thereby recognized the value of maritime clusters as an instrument and stated that the Commission would encourage the formation of multi-sectoral clusters and would encourage a European network of maritime clusters. The IMP also promotes the application of integrated approaches to maritime policy at the level of member states, who agreed to provide information on national approaches and on maritime cluster initiatives. The EC continued to support cluster activities with general policies such as 'Towards world-class clusters in the European Union: Implementing the broad-based innovation strategy' (2008b).

Main characteristics of maritime cluster organizations

Maritime cluster organizations aim to enhance the strength of the maritime cluster. By definition, maritime cluster organizations capture more than one maritime sector. Besides cluster organizations that provide a platform for all companies in sectors that are related to each other, sector associations link all companies and/or organizations within a specific sector. Consequently, cluster organizations and sector associations complement each other. In this respect, sector associations often join a cluster organization to team up with other sector associations and companies in order to cooperate on the realization of common interests.

An analysis of representation of maritime cluster organizations in European star-regions indicates the main similarities and characteristics of these cluster organizations (Monteiro et al. 2013). Most European maritime star-regions are represented by a national and/or a regional cluster organization. This indicates that most European countries with a star-region in the sea-related sectors have access to those sectors through a maritime cluster organization. There are basically two possible approaches for initiating a cluster: initiatives in government-induced organizations (top–down) and cluster organizations induced by leader firms and/or sector associations (bottom–up).

The main goal in establishing cluster organizations is the installation of a structure that provides clear solutions for cluster issues (e.g. labour market and innovation), regardless of the initiator of that cluster organization. The main aims are: to increase competitiveness, to promote maritime sectors, and to improve coordination within the cluster. The organizational structure depends on different climates and/or environments, influenced by traditions, history, culture, governmental structure and driving sectors. Basically, cluster issues need to be handled within the cluster organizations based on the following key elements:

- Transparency and communication/dialogue, both horizontally and vertically, on the mission, structure and activities of the cluster organization.
- The presence of leading individuals within the cluster organization who can raise enthusiasm and set up clear actions in view of a clear and supported vision.

In terms of scope, maritime clusters cover different geographical areas. The focus of the organizations can be on an international and/or European level, a national or regional (including cross-border) level, or a local level. A clear distinction between the main characteristics and activities of national and regional maritime cluster organizations can be made. Next to their geographical scope, these organizations show differences and similarities in several areas.

A first difference is that national cluster organizations, in general, were established in an earlier stage than regional organizations. Also, their purposes differ. While national organizations focus more on establishing a strategic framework, regional organizations focus more on improving local development and operations. Thus, the main topics covered by national clusters concern the labour market and the exchange of know-how as regards to innovation, research and image building; whereas regional clusters mainly cover topics such as local business development. Another difference between the two cluster organizations is the higher budgets of national cluster organizations to finance activities and initiatives.

The establishment of national and regional maritime cluster organizations is driven by a similar rationale and these organizations have almost similar types of members that are active in the cluster organization. Except for national top–down cluster organizations, companies and sector associations are involved as members in the activities of the cluster organization.

Organizational structure

Besides good practices of maritime cluster organizations based upon the type of activities they carry out, cluster organizations can optimize cluster benefits through their organizational structure. In order to optimize efficiency, to increase the level of business formations and to reach a higher level of innovation, this organizational structure should include all stakeholders that are logically linked to each other, be it as an initiator or as a member of that organization. These stakeholders should have an equal weight in the decision-making process within the cluster organization. The logically linked stakeholders are governmental bodies, sectors represented by trade associations, and private persons. For example, the sectors of shipbuilding and shipping are logically related to each other: shipbuilding is the supplier of shipping. Sectors that have limited logical links and/or relations with other maritime sectors experience limited cluster benefits.

A public–private integrated cluster organization captures all these maritime stakeholders in order to have well-structured discussion between the government and the sector. Other important benefits of this public–private integrated maritime cluster organization are:

- A cluster-wide approach.
- Focus on long-term growth and development as well as on (often short-term) sectors' interests.
- The ability to create a level playing field through legislation.
- Sector involvement in the evaluation of the cluster organization (through membership contribution).
- Improved communication both horizontally and vertically.

SWOT analysis of maritime cluster organizations

A SWOT analysis of maritime cluster organizations distinguishes the strengths, weaknesses, opportunities and threats of both top–down and bottom–up clusters.

The main strength of top–down maritime cluster organizations is that they are initiated by the government and consequently aim to increase the growth of all maritime sectors within the cluster. This growth is supported by a long-term strategy and policy, focusing on the improvement of research, development and innovation, and the creation of a level playing field through legislation within the maritime cluster. Through research, development and innovation (RDI) and legislation, as well as relatively high budgets, top–down clusters can positively influence common interests (e.g. the environment). However, one of the identified weaknesses of top–down cluster organizations are the difficulties they experience in defining to which cluster(s) a sector belongs. This occurs, for example, in categorizing ports. Ports belong both to the logistics cluster and to the maritime cluster. Moreover, it is not always clear which interests are at stake or how they can be aligned with other (maritime) sector interests.

Another challenge of top–down cluster organizations is caused by the continuous balancing process between the interests of cluster organizations and those of sector associations, requiring constant interaction. As a result of the widespread variety of interests, this often leads to a focus on more universal, softer themes. In this respect, the surplus value of the maritime cluster organizations needs to be proved constantly. Opportunities for top–down cluster organizations are the increasing cooperation on European level to discuss best practices and lessons learned. A second opportunity is that cluster organizations could take the initiative (or be invited) to establish a single point-of-entry to increase structural interactions. The organizational structure of the maritime cluster in the Netherlands is an example of this public–private integrated maritime cluster.

Whilst there is an opportunity for top–down clusters to increasingly involve sectors' opinions in the structural evaluation of cluster organizations' activities, the limited in-depth cooperation between the ministries involved in the maritime sectors could be a possible threat. This limited cooperation could lead to a suboptimal integrated maritime cluster policy. Another potential threat is that top–down maritime clusters have limited structural sector involvement. This may lead to focusing too much on long-term benefits for the maritime sectors that may hamper or conflict with the clusters companies' short-term interests.

As regards bottom–up cluster organizations, the main strength is the enhancement of business efficiency and opportunities. This strength goes hand-in-hand with a strong focus on finding solutions for operational problems. Moreover, cluster organizations are structurally evaluated by their members with implications for improved membership contributions. A third strength is the ability of the cluster to operate as a single point-of-entry for the government to discuss sectors' interests (e.g. tax and regulation). Also, bottom–up cluster organizations often take the lead in organizing events (e.g. to promote working in the cluster) and in establishing issues platforms (e.g. to exchange best practices and lessons learned).

On the other hand, a weakness of bottom–up clusters concerns the differences in sector interests, sometimes leading to a difficult decision-making process within the cluster organizations. This difference in sector interests often results in a rather narrow focus on softer themes (e.g. promotion). Furthermore, company and sector interests in bottom–up clusters are often not balanced. This potential imbalance is sometimes caused by differences in the weight and influence of members or supporting companies and sectors within the cluster organizations. In particular, this imbalance is often caused by the strong impact of the largest contributing members in the decision-making process. Finally, another weakness is that bottom–up cluster organizations are sometimes not always much focused on strategy and long-term growth of the sectors involved and on the cluster as a whole.

One of the opportunities for bottom–up clusters is the increasing cooperation at European level to discuss best practices and lessons learned. A second opportunity is that bottom–up cluster organizations take the initiative to increase the government involvement (and budget) in the cluster organization to improve mutual understanding and to fine-tune policy making. While another opportunity for bottom–up cluster organizations is the focus on longer term maritime sector benefits like growth and strategy, a main threat for cluster organizations is that a too narrow local focus may lead to the suboptimal functioning of a maritime cluster organization. Also, the limited attention for the role of (relatively) smaller sectors and companies within the maritime cluster poses a threat to the bottom–up cluster organizations.

European Network of Maritime Clusters

The European Network of Maritime Clusters (ENMC) is a confederation of clusters or equivalent structures. The ENMC is regarded as a best practices dissemination and exchange platform through their website, informal talks and an annual summit during which each country gives a brief presentation of the economic situation of its maritime sector and the recent actions of its national organization. The aim is to establish a framework for future common targeted actions. All of the member organizations are, or tend to be, cross-industry organizations gathering all or part of the maritime subsectors of their countries. The type of each national cluster varies, with some being almost state-controlled, some purely private-owned and some being an intermediate mix. Some clusters include inland navigation and/or logistic sectors, port industries, coastal tourism, insurance and finance in their scope.

Under the responsibility of a presidium, and consisting of a chairman, a vice-chairman and a secretary-general, the ENMC has grown from 10 countries in 2005 to 18 countries in 2015, and has continuously extended its membership and geographical coverage. The members of the ENMC regard eight sectors as part of the maritime cluster from a narrow perspective: shipping, shipbuilding, marine equipment, seaports, maritime services, yacht building, offshore services and fishing. Three sectors can be added from a national, broader perspective: navy and coastguard, inland navigation and maritime works.

Maritime cluster organizations cooperate at different levels. The ENMC aids transnational cooperation between national organizations on an EU level, although bilateral contacts with other national and regional maritime cluster organizations and with sector associations are also quite frequent. Regional maritime cluster organizations mainly cooperate with other regional and national cluster and sector associations. The main topics and activities of this cooperation are meetings/events, data and information sharing, exchange of lessons learned and best practices, policy initiatives, projects/studies, and publicity and promotion.

Cluster benefits in theory and practice

Clustering is not a goal in itself, but may be beneficial because of its link with economic prosperity. Regions with a higher rate of employment in industries that belong to strong clusters appear generally more prosperous (i.e. a higher GDP per capita). Other key aspects and benefits of clustering are related to:

- *Innovation:*
  - clusters provide a fertile environment for innovation;
  - clusters are a vital element of the Lisbon strategy;

  - cluster firms benefit from the flows of tacit knowledge and the presence of skilled labour;
  - cluster firms are more innovative than non-cluster firms;
  - cluster firms register more trademarks (29% vs. 14%) and apply for more patents (29% vs.12%) than non-cluster firms; and
  - clusters are well aligned with the modern approach of 'open innovation' (i.e. innovation is created in open environments rather than by isolated organizations).

- *Specialization:*

  - of all European employees, 38% work in enterprises that are part of a cluster;
  - individual cluster categories differ significantly in the level of employment concentration; and
  - in terms of cluster strength, Europe lags behind the US (probably due to the remaining barriers to cross-regional competition in Europe and the differences in entrepreneurial cultures).

- *Outsourcing:*

  - cluster firms contract out more research (41% vs. 20%) and carry out less research in-house (44% vs. 53%).

The main theoretic benefits of cluster organizations' activities can be divided into an increase in efficiency, an increase in level of business formations, and a higher level of innovation. Companies within a cluster can operate with a higher level of efficiency by drawing on more specialized assets and suppliers with shorter reaction times than they would be able to in isolation. Also, the level of business formations tends to be higher in clusters. Start-ups are more reliant on external suppliers and partners, all of which they find in a cluster. Clusters therefore reduce the costs of failure, as entrepreneurs can fall back on local employment opportunities in the many other companies in the same field. Finally, companies and research institutions can achieve higher levels of innovation. Knowledge spillovers and the close interaction with customers and other companies create more new ideas and provide intense pressure to innovate while the cluster environment lowers the cost of experimenting.

Activities and initiatives of clusters

The theoretic benefits of a cluster can be achieved through the activities of maritime cluster organizations. Firstly, an increase in internal and external communication with regard to the maritime clusters through promotion can enhance possibilities to cooperate efficiently. Secondly, by focusing on education, training and the labour market, the level of business formations can be increased. Thirdly, a focus on research and development contribute to obtaining a higher level of innovation.

The link between cluster policies and sector performance

There are basically three types of cluster policies:

- *Facilitating policies:* Create a favourable microeconomic business environment for growth and innovation (indirect stimulation of the emergence and dynamics of clusters).
- *Traditional framework policies:* Include industry and SME policies, research and innovation policies, and regional policies.
- *Development policies:* Aimed at creating, mobilizing or strengthening a particular cluster category resulting in specific sectoral cluster initiatives.

In particular, development policies focus specifically on clusters and clustering. This policy focus is welcomed by the private sector, since over 68% of companies working in a cluster-like environment agree that public authorities have a fundamental or important role to play in supporting the cluster. Despite the fact that clustering and cluster policy are currently receiving significant attention, the cluster policy area is still at an early stage. Reasons for this are that the use of cluster policy only started in the period 1990–1994 and that around half of the countries using cluster policy, did so for the first time in the period from the year 2000 until today.

The effect of cluster policies on sector performance has, up until today, proven very difficult to measure. One of the reasons for this difficulty is the early stage of development of cluster policies, but another is the difficulty in measuring the impact of cluster policies and programs as most of their effects are indirect and linked to many other factors. These elements make it difficult to establish clear causal links between cluster policies and programs and their (potential) impact. Moreover, only very few cluster organizations (including the maritime sectors) measure or assess the impact of their activities on the functioning of their clusters. Nevertheless, cluster organizations are structurally evaluated by their members with implications on membership contributions.

A cluster is something that is actively, purposefully created with the commitment of key players, institutions and the nation as a whole. It cannot be enforced, but only negotiated as initiatives are encouraged and supported to cooperate in multiple spheres such as innovation, export forums, promotion and recruitment, and policy development. The benefits of a cluster will be far higher than the sum of the activities of its individual members.

European cluster networks: Summary and recommendations

The conclusions on the economic impact of the sea-related sectors and clusters in Europe are the following:

- The maritime sectors in Europe account for about 4.8 million persons employed and turnover is about €450 billion.

- The maritime sectors generate an added value of circa €186.8 billion, of which 66% is in the traditional maritime sectors, 25% in coastal and sea-related (marine) tourism and recreation and 9% in fisheries.
- The largest maritime sectors in terms of added value are shipping, seaports, marine equipment, navy and shipbuilding.
- Star-regions in the traditional maritime sectors are mostly spread over north-western Europe; for coastal (and marine) tourism and recreation, star-regions are located more in southern Europe; star-regions for fisheries are spread over all European regions.
- In addition to its explicit economic value, the maritime clusters in the EU also play a key role in facilitating the functioning of the entire economy, for example by means of maritime transport facilitating international trade, and by generating indirect effects through purchases in the value chain.

Based upon the aforementioned characteristics, the main benefits and the SWOT analysis of these organizations, the main conclusions of the role of maritime cluster organizations are the following:

- Maritime companies and sectors in all European countries with a maritime cluster have access to that cluster through a maritime cluster organization.
- The main aims for the establishment of a maritime cluster organization are to increase competitiveness, to promote maritime sectors, and to improve coordination within the cluster.
- Cluster issues need to be handled within the cluster organizations based on transparency, communication/dialogue and on the presence of leading individuals within the cluster organization.
- Policy (initiatives) and actions are translated to European, national and regional levels, although not in a uniform manner.
- Maritime cluster organizations are mostly historically established (and regionally grown), based upon the specific needs of the cluster and its participants and upon the country's culture.
- In general, top–down and bottom–up initiated clusters focus on the same topics; and the main topics are promotion of the maritime cluster, education, training and the labour market, and focus on RDI.
- Although the impact/success of actions of cluster organizations is only limitedly evaluated on a project basis, these organizations are indirectly evaluated through (the continuation of) membership contributions.
- There is no organizational concept of cluster organizations that intrinsically excels, because much depends on the local conditions (e.g. presence of sectors and regional spread of sectors) and the practical elaboration of activities and topics. Topics of cluster organizations concern mostly softer issues (e.g. innovation and labour market). In top–down maritime cluster organizations, more opportunities are created for strategic issues.
- Cluster organizations and sector associations (most often) will complement each other.

At an ENMC meeting in Rome, the cluster organizations recognized the added value of the cooperation among clusters on a European level. This recognition was confirmed during the workshop with representatives from maritime clusters, government officials and experts in this area on 6 October 2008, during which the draft results of the current study were presented. In terms of recommendations, the following suggestions were made:

- A continuous effort to build a permanent database on economic facts and figures to work towards a monitoring system in which national cluster organizations play a key role. A round table (of ministers) could be organized to discuss and launch the initiative.
- A clear-cut approach from the EC with respect to maritime clusters. This implicates that the importance and the strength of maritime clusters is well communicated, for example by integrating the results of this study in the website of the European Cluster Observatory.
- A strategy towards a future role of European, national and regional maritime cluster organizations. A platform could be installed to exchange best practices in the field of labour markets. Also, where possible, integration of marine and maritime affairs is recommended. Methods to strengthen RDI need further investigation. Finally, cluster organizations should take a leading role in environmental protection and sustainability.
- In case the government does not take the initiative to set up a cluster organization, or does not provide incentives to do so, the industry needs to take this up on its own initiative if it wishes to set up coordinated actions. The main goal in establishing a maritime cluster organization does not so much depend on the initiator, but on the installation of a structure that provides clear solutions for cluster issues (e.g. labour market and innovation) using key elements such as transparency, communication and the presence of leading individuals. This transparency of the functioning of cluster organizations' activities is emphasized and supported by the initiatives of the European Cluster Observatory; enterprises can use the Observatory's information and communication on cluster organizations to their advantage in order to benefit from being actively involved in clusters.

CASE STUDY: CENTRAL BALTIC REGION

Introduction

The maritime clusters in the Baltic Sea region are facing various new challenges. The competitive advantage of the Far Eastern maritime shipbuilders lies in producing series of standardized vessels at low costs, whereas the European shipyards, with their subcontractors and suppliers, provide specialized solutions and technologies. For the European maritime clusters, keeping one step ahead of the fast-growing Asian competitors is not an easy task. At the same time, the local environment is facing changes that affect shipping companies, ports and port operators, including tightening environmental regulations as well as rising cost levels. The political decision makers, along with various interest groups and associations, also strongly affect the maritime actors' playground. All these developments have an impact on cluster competitiveness; so what is the future outlook for the Baltic Sea region?

The importance of cooperation

Smart Competitiveness for the Central Baltic region (Smart Comp) is a Central Baltic-financed project which aims to support smart, environmentally sustainable development, growth, competition and cooperation between maritime clusters, cities and universities in the Central Baltic region (CBR: Estonia, Finland, Latvia and Sweden). In the midst of common challenges, how can the maritime clusters in the CBR improve their competitiveness? Although the maritime clusters of the CBR have several factors in common, they are still separate and often compete with each other. However, in the future they should find ways to combine their strengths in order to increase the competitiveness of the region's maritime sector.

The risk of accidents, such as oil catastrophes, concerns all coastal states, and boosts the investments in technologies and processes aimed at increasing the safety of navigation. The legal framework for vessel traffic in the Baltic Sea is also tightening, for example the sulphur emission regulations. The maritime clusters in the CBR have to make remarkable investments in finding and implementing new solutions to meet the regulations. Although challenging, the tightening environmental regulations could also turn into an opportunity, making CBR countries forerunners in environmentally friendly technologies, life-cycle solutions and fuel efficiency. At the same time, the development of the Russian ports can increase material flows in the region, and the CBR clusters have to make sure they are involved and benefitting from the Russian business opportunities.

With rising cost levels threatening the profitability of the maritime industry, and global competition tightening as Asian clusters compete with lower production costs, the CBR maritime clusters need to compete with quality and specialization, rather than concentrating on price. This requires maintaining sufficient financial and scientific resources for innovation activities. Availability of qualified workforce is also a common problem in the maritime sector, both due to aging population and to experts shifting to other industries. This challenge has to be met by guaranteeing the sufficiency and quality of education and creating an appealing image for the maritime sector. Common branding and communication activities could improve the image of the maritime sector and increase awareness of its importance for the CBR countries.

The relatively small size of the clusters in the CBR mean that common challenges could be better answered with coordinated solutions. The cluster companies should not only strengthen their internal cooperation networks but also establish connections to other clusters abroad in order to increase their business opportunities, knowledge flows and joint research and development (R&D). Encouraging this, however, requires government support for SME internationalization and innovation activities as well as new mechanisms to finance investments. However, the cluster competitiveness ultimately stems from companies and their efforts for continuous development. By viewing each other less as competitors and more as cooperative partners and by joining their forces around the Baltic Sea, the maritime clusters in CBR could better tackle their common challenges and compete with their large Asian counterparts.

Clusters and competitiveness

Today, competition does not take place only within companies but also increasingly within networks. The competitive advantage of a company is largely determined by the characteristics of the network to which the company belongs, and by the way it organizes the activities and uses the resources in its value chain relative to its competitors. Due to the global nature of markets, specialization and innovation are essential in maintaining competitive advantage. In this process, the establishment of networks is a quintessence (Álvarez et al. 2009), allowing companies to focus on and develop their core activities. A cluster is a specific type of network: a geographic agglomeration of companies that are vertically and horizontally linked by channels for business transactions, cooperation and/or competition. They share a localized support infrastructure, labor markets and services, and are faced with common market opportunities and threats.

While global competition continues to drive clusters towards higher levels of specialization and innovation (Porter 2008), it is widely accepted that both competition and cooperation relationships may contribute to the development and competitiveness of the involved cluster companies. Geographical proximity facilitates mutual visibility and transparency among the cluster companies (also competitors) by which companies are able to observe each other's competitive developments and improve their own operations. Thus, the competitiveness of the local industrial cluster increases (McCann 2008).

Clusters are of value also in terms of knowledge creation and innovation cooperation. Within a local cluster there is a thick web of information, knowledge and inspiration which circulates between the actors of a cluster. This 'local buzz' consists of information flows, knowledge transfer and continuous updates as well as of learning opportunities in both organized and spontaneous meetings. Supra-regional linkages or 'pipelines' are of great importance. Through such pipelines to other clusters, new information about market developments and technologies is pumped into the local cluster and its networks, further stimulating the local buzz. However, building access to trans-local pipelines requires considerable efforts, commitment, time and money. The interaction depends on trust between the partners, who, at the same time, should be on the same 'wavelength' and share context that is similar enough to enable joint problem solving and knowledge creation.

Through such strategic partnerships of inter-regional and international reach, cluster firms may continuously combine and re-combine similar and non-similar resources to produce new knowledge and innovations. Fruitful inter-cluster environment for innovation generation is of great importance in the maritime industry, where retaining competitive edge requires continuous innovations in both product and process. However, it must be noted that competitiveness of a company is never solely a matter of its networks or cluster capabilities. The business environment where this agglomeration operates has a significant impact on its development by determining the opportunities for a company to increase its capabilities, to make competitive choices and to generate output from its assets.

Reference to the surrounding business environment, however, means that almost everything eventually matters for competitiveness and no single policy, for instance, can create competitiveness (Porter 2008). Nevertheless, through governmental policy strategies, the cooperation among local cluster companies as well as with external networks can be promoted, providing the cluster firms with a greater direct accessibility to complementary firms and sectors, R&D activities and innovation systems. Furthermore, in any given cluster at a particular time, there are elements representing the key barriers that restrain companies from reaching higher levels of productivity. If these key barriers remain unsolved, improvements in some other areas of the business environment may have little or even negative effects on productivity (Porter 2008). Thus, identifying and tackling the main bottlenecks provides the most efficient way to support the competitiveness of a given cluster.

Regarding the issues that affect the competitive advantage of a cluster, the economic business environment comprises four main groups of factors: (1) the context for firm strategy and rivalry, (2) input conditions, (3) demand conditions, and (4) the related and supporting industries. This framework is often referred to as the competitive diamond. Based on the four facets of the diamond, the value creation potential of clusters stems from the environment in which they operate. All the diamond factors are important in creating a well-functioning industrial cluster, but the influencing of these factors resides from the fact that they are all highly interconnected. While the role of private sector in cluster upgrading is present in all parts of the diamond, effective solutions on improving the local business environment often require governmental actions (Porter 2000).

Therefore, in the case of the maritime cluster in the Baltic Sea region, the way to increase its competitiveness derives from identifying the key challenges related to this particular industry network and location. Although Porter's cluster theory (1990) has faced criticism for its profundity (for example, Martin and Sunley 2003), it has proven by far the most influential tool for cluster environment analysis and has become a standard concept in the field.

Key competences of each cluster in the region

The CBR maritime clusters have several factors in common but there are also country-specific strengths. Regarding Estonia, tourism, transport and logistics are important elements of the maritime cluster as the majority of foreign trade as well as international traveling is seaborne. The Tallink Group dominates the shipping business, particularly regarding passenger traffic, but also provides Ro-Ro cargo services. Transit trade, particularly from and to Russia, is important for Estonia and its ports. The majority of goods handled in Estonian ports are petroleum products which are transported from East to West through Estonia. In the shipbuilding sector, ship repair and maintenance is well developed and the Baltic Ship Repair Company (BLRT) dominates the industry even throughout the Baltic States. The fishing industry is also well developed in Estonia.

In Finland, the shipbuilding competence is extensive and shipbuilding is one of the most important sectors of the country's maritime cluster. Shipbuilding, together with a wide network of related and supporting industries specialized particularly in the needs of the Finnish maritime industry, can be said to form a cluster of its own. Finland is also an important player in the offshore sector. There is a demand for excellent quality as well as special niche know-how and technology, for instance related to the Arctic conditions. As around 85% of the Finnish foreign trade is seaborne, shipping industry and ports are an important part of the Finnish maritime cluster. In addition, Russian trans-shipments are of great significance for Finnish ports, where they are competing with ports in the Baltic states.

Regarding Latvia, transports and logistics play an important role in the national economy and ports are an important element of the country's logistic cluster, as well as the maritime cluster. The competitiveness of ports is largely affected by railway and road accessibility, port infrastructure and various services offered in the vicinity of the ports. Trans-shipments from Russia and other Commonwealth of Independent States countries constitute the majority of the cargo transported through Latvian ports. Latvia's shipbuilding industry is mainly specialized in ship repair activities. There is also a wide and competitive network of suppliers of marine industry, whose products range from ship hulls and machinery to ship design, interior and various electronic devices.

The Swedish shipbuilding sector went through a structural change some decades ago. As a consequence, all large shipyards were shut down and currently there are only a number of small shipyards operating in Sweden, specializing mainly on repair work. However, there is still a wide and competitive network of various suppliers of marine industry which represent industrial, technology, engineering, and design companies with a particular specialization in the offshore and steel industries. In addition, the building of pleasure boats has developed into an important field of operations for the Swedish maritime cluster.

What kind of actions should be taken?

The situation at hand is challenging, in that profitability is low for many companies, the market is scattered and there is fierce competition from other countries both near and far. Therefore the current trend is managing costs, not revenues. Environmental regulations push for change, but few stakeholders have acted on them. The only way to stay competitive in the long term is to invest in innovative solutions that bring added value to the end customer. This relates to shipyards, shipowners, ports and suppliers. Developing the optimal logistical solutions demands a higher degree of cooperation between the actors in the respective clusters as the whole value chain is needed to formulate them. Today many industrial companies are doing this work themselves as they have not found a service provider who can offer a competitive enough solution.

Current fleets need to be reviewed and renewed, which means getting rid of old tonnage and making sure that the rest meets the new demands regarding environment and fuel efficiency. The resulting retrofits, conversions and new buildings will provide business opportunities for the shipyards in the region. The shipyards and suppliers need to consider the life-cycle aspects more closely at the design stage, and offer these services as part of their portfolio. In order to create new ship designs that ensure competitiveness in the future, joint R&D is needed which could bring shipowners, end customers, shipyards and designers to the same table. As the Baltic Sea belongs to SECA (the Baltic Sea Sulphur Emission Control Area), they are the first to implement lower sulphur requirements. This provides them with the opportunity to devise a good solution which can then be exported to other regions that are following the requirements 5 years later.

Innovative fuel solutions, taking into account the ecosystem, need to be developed. In addition to fuel efficiency, ship designs optimizing the usage of space and cargo handling (to minimize damages and the time needed for loading and discharging) are needed. In order to secure financing for the necessary investments, new funding mechanisms should be created. Traditional bank financing is hard to obtain because some major ship financing banks have ceased business due to poor profitability. In addition to banks, shipping funds are needed, and the right models for them have to be created. The role of the governments should not be forgotten: the shipbuilding and shipping industries have been supported by the local governments in different ways, such as subsidies in the form of tax exemptions or receiving innovation support.

Conclusion

In the future, the CBR maritime cluster will need the support of the government, but this support should not be provided automatically. Rather it should be connected to the industry's own efforts in improving its competitiveness and demonstrating initiative to develop innovative solutions to the benefit of the entire cluster and region or country. Indeed, a higher degree of collaboration is needed both within as well as between the clusters. Facilitating the networking among companies and combining their resources could increase their competitiveness and improve their chances of penetrating international markets, for instance Russia. Furthermore, free flow of ideas within the clusters can support the innovation development, and common challenges often require coordinated solutions. Viewing each other less as competitors and more as cooperative partners could benefit the maritime clusters, and joining their forces could bring competitive advantages for the whole region.

Discussion

The concept of clusters has been widely integrated into the conduct and political governance of maritime activities, both at the level of the EU and in the respective member states. This development can be seen as evidence of the concept's value, and in addition, it becomes even more worthwhile to familiarize oneself with the basic tenets of clusters, their structure, and their economic significance. Not only do clusters exist, they are highly significant to European economies, as regions with higher levels of clustering tend to have higher income and employment levels. The total economic impacts of maritime clusters are of great importance to most European nations, as they consistently account for multiple percentages of total GDP and employment.

References and further reading

Álvarez, I., Marin, R. and Fonría, A. (2009). 'The role of networking in the competitiveness of firms'. *Technological Forecasting & Social Change*, 79: 410–421.

Community Attributes Inc. (2013). 'Washington State maritime cluster: Economic impact study'. www.maritimefederation.com/uploads/2/9/9/6/29962189/cai.wa-maritime-cluster-study.2013-1120.pdf (accessed: April 2017).

European Commission (EC) (2008a). 'The role of maritime clusters to enhance the strength and the development of the European maritime sectors'. https://ec.europa.eu/maritimeaffairs/sites/maritimeaffairs/files/docs/body/report_results_en.pdf (accessed: April 2017).

European Commission (EC) (2008b). 'Towards world-class clusters in the European Union: Implementing the broad-based innovation strategy'. http://eur-lex.europa.eu/legal-content/EN/TXT/?uri=celex%3A52008DC0652R(01) (accessed: April 2017).

Finpro (2012). 'East Asia value networks – Case maritime cluster'. www.finpro.fi/documents/10304/77620/TF_East_Asian_Value_Networks_Maritime_Cluster.pdf (accessed: April 2017).

Kildow, J.T. and McIlgorm, A. (2010). 'The importance of estimating the contribution of the oceans to national economies'. *Marine Policy*, 34: 367–374.

Kraaijeveld, A. (2012). 'Cooperation is key for the Dutch maritime industry.' *Maritime by Holland*, 61(3): 24–26.

Laaksonen, E. and Mäkinen, H. (2013). 'The competitiveness of the maritime clusters in the Baltic Sea region: Key challenges from the Finnish perspective'. *Journal of East-West Business*, 19: 91–104.

Makkonen, T., Inkinen, T. and Saarni, J. (2013). 'Innovation types in the Finnish maritime cluster'. *WMU Journal Maritime Affairs*, 12: 1–15.

Martin, R. and Sunley, P. (2003). 'Deconstructing clusters: Chaotic concept or policy panacea?' *Journal of Economic Geography*, 3(1): 5–35.

McCann, P. (2008). 'Agglomeration economics' in: C. Karlsson (ed), *Handbook of Research on Cluster Theory*. Cheltenham, UK: Edgar Elgar, p.23–38.

McKinley, E. (2012). 'Future collaboration for the marine sector in Devon: Pulling together'. http://eprints.chi.ac.uk/1742/ (accessed: April 2017).

McKinley, E. and Fletcher, S. (2010). 'Individual responsibility for the oceans: An evaluation of marine citizenship by UK marine practitioners'. *Ocean & Coastal Management*, 53(7): 379–384.

McKinley, E. and Fletcher, S. (2012). 'Improving marine environmental health through marine citizenship: A call for debate'. *Marine Policy*, 36(3): 839–843.

McKinley, E., Robins, D. and Cooper, D. (2012). 'Maritime clusters: A governance success? Evidence from the CAMIS Project'. www.vliz.be/imisdocs/publications/241232.pdf (accessed: April 2017).

Monteiro, P., Noronha, de T. and Neto, P. (2013). 'A differentiation framework for maritime clusters: Comparisons across Europe'. www.mdpi.com/2071-1050/5/9/4076 (accessed: May 2017).

Policy Research Corporation (2001). 'Economic impact of maritime industries in Europe'. Final report. Antwerpen: PRC.

Porter, M.E. (1990). *The Competitive Advantage of Nations*. New York: Free Press. (Republished with a new introduction, 1998.)

Porter, M.E. (2000). 'Location, competition and economic development: Local clusters in a global economy'. *Economic Development Quarterly*, 14(1): 15–34.

Porter, M.E. (2003). 'The economic performance of regions'. *Regional Studies*, 37(6&7): 549–578.

Porter, M.E. (2008). 'The five competitive forces that shape strategy'. *Harvard Business Review*, 86(1): 78–93, 137.

Smart Comp (2013). 'Maritime companies and their business networks in the Central Baltic region'. Research Report No.2. www.utu.fi/fi/yksikot/tse/yksikot/PEI/tutkimus/Documents/SmartComp%20Research%20Report%20June%202013.pdf (accessed: April 2017).

Stavroulakis, P.J., Papadimitriou, S. and Koliousis, Y. (2015). 'The competitive advantage of maritime clusters'. www.researchgate.net/publication/277330083_The_Competitive_Advantage_of_Maritime_Clusters (accessed: April 2017).

Chapter 16

Clusters and stewardship

> Water and air, the two essential fluids on which all life depends, have become global garbage cans.
>
> Jacques Yves Cousteau

Introduction

More than half of the world's population lives within 60 km of a coastline. By the year 2020, that figure will have risen to three-quarters. The world's coastal areas are thus exposed to rapid urban growth, increasing population strains, pressures from major industries, particularly tourism, and extensive exploitation of marine resources. The results include the pollution of marine and freshwater resources, air pollution, loss and degradation of marine and land resources, the depletion of cultural resources, loss of public access, and increasing levels of noise and congestion.

Tourism is the most important industry in many coastal areas. By 1994, international world tourist receipts had reached more than US$320,000 million a year and the number of international tourists totalled 528 million. Much of this money and many of these arrivals affect coastal areas since the beach holiday 'sun and sand' tourism is still the most popular form of tourism. Coastal areas are thus one of the tourism industry's principle resources, and there is an urgent need to protect and improve coastal resources.

Although there have been many attempts to protect coastal areas and to encourage sustainable forms of coastal development, few have been successful. The main reason for this is that they have largely been sectoral, and there is fierce competition for coastal resources in many areas. For example, there is often conflict in coastal areas over access to the coastline, which is required for tourist beaches, marinas, aquaculture and cooling for power generation.

Over the years, wastewater has been a source of pollution due to urbanization, growing cities, industrialization and improved material consumption, among other factors. Today, an estimated 80% of global wastewater is being discharged untreated into the world's waterways. This affects the biological diversity of aquatic ecosystems and disrupts the fundamental web of our life-support systems, on which a wide range of sectors (from urban development to food production and industry) depend. Yet, if properly managed, wastewater could be a source of water, energy, fertilizer and other valuable materials and services. Each year, for instance, approximately 330km^3 of municipal wastewater is generated globally. A recent study showed that resources embedded in this wastewater would be enough to irrigate and fertilize millions of hectares of crops and produce biogas that could supply energy for millions of households (UNEP 2015a).

Nations that promote their coastal areas for tourism are increasingly becoming aware of the need to protect these areas in order to maintain their natural beauty and help ensure their long-term vitality as tourism destinations. Tourists today have a greater understanding of the impacts of their travel and are demanding more sustainable tourism products. In essence, the majority of tourists today are looking to satisfy their need for leisure, recreation and discovery in a way that is friendly to the natural, cultural and social wellbeing of the destinations they visit. In response to the growing need for coastal zone protection, the tourism sector has been working to develop approaches and strategies that allow for the better planning and management of tourism activities in coastal zones. However, this has not always been an easy task as these extremely fragile environments are targets for many other human-induced development activities (UNEP 2009).

Emerging trends in conservation and protected areas management set the stage for new approaches that engage local people in the stewardship of working seascapes that embrace the interactions of people and nature. One trend is that conservation strategies are becoming increasingly bioregional. The field of conservation biology has highlighted the pressing need to work on the scale of ecosystems and the wider seascape to conserve biological diversity.

Another important change lies in how we view land-based national parks and marine protected areas. Worldwide, there is growing recognition that protected areas can no longer be treated as islands, but must be seen in a larger context. In regions such as Latin America and the Caribbean the phenomenon of 'paper parks', protected areas in name only, has demonstrated forcefully that approaches that rely solely on regulation and enforcement are costly and too often meet with failure. Protected area managers are turning instead to 'inclusive' models, in which the interests of local communities are considered, resident populations are not displaced, and there is a high degree of local participation in planning and management of the protected area.

A third trend lies in our growing understanding of the link between nature and culture: that healthy seascapes are shaped by human culture as well as the forces of nature, that rich biological diversity often coincides with cultural diversity, and that conservation cannot be undertaken without the involvement of those people closest to the resources. Fundamental to these new directions in protected areas is the need to engage and support local people in the stewardship of their natural and cultural heritage.

The stewardship approach

Stewardship means, simply, people taking care of the Earth. In its broadest sense, it refers to the essential role individuals and communities play in the careful management of our common natural and cultural wealth, both now and for future generations. More specifically, it can be defined as efforts to create, nurture, and enable responsibility in stakeholders and resource users to manage and protect the oceans and its natural and cultural heritage.

Stewardship taps our basic human impulse to care for our home and its surroundings, be it a parcel of land, a neighbourhood or a historic monument, or the larger area of a watershed, mountain range or stretch of coastline. It builds on our sense of obligation to other people: our family, our community and future generations. By fostering individual and community responsibility, the stewardship approach puts conservation in the hands of the people most affected by it. Stewardship emphasizes the integration of people and nature, not the attempted isolation of one from the other. It recognizes that all seascapes are cultural, and that conservation needs can be addressed on oceans and seas that cannot be removed from human existence and commerce. Seascapes typically encompass a mosaic of different users and land ownership practices: private, public and, in many countries, customary or communal ownership.

The scenic, biological and cultural qualities that make certain landscapes and seascapes unique are the result of the interactions of people and nature over time. It follows that protection of these landscapes and seascapes inevitably must rely on fostering stewardship by those who own or live on the land in coastal areas. Experience with private land stewardship in North America, and increasingly, in other regions of the world, offers an array of tools to conserve the natural and cultural values of landscapes and seascapes.

Specific stewardship tools vary according to social, legal, ecological and institutional constraints, but all operate to encourage, enable, or formalize responsible management. Briefly, these techniques include environmental education, technical information, demonstration projects, recognition of achievement, certification, voluntary management agreements, subsidized management, deed restrictions, public–private partnerships in protected areas management, and outright acquisition of property by private organizations. These tools (with many others and more variations) represent a spectrum of options beginning with those that require little or no formal commitment or involvement and little per capita investment (e.g. education), to more 'permanent' and specific protections (e.g. easements and acquisition).

While national parks and other strictly protected areas are essential, they alone cannot achieve biodiversity conservation objectives, nor can they encompass working landscapes and seascapes. There is therefore a pressing need for new models of protected areas. Beresford and Phillips (2000) claim that the protected landscape approach is central to a new paradigm for protected areas, one which is based on inclusive approaches, partnerships and linkages. This approach can provide valuable models for integrating biodiversity conservation, cultural heritage protection and sustainable use of resources.

According to the International Union for Conservation of Nature (IUCN: see Philips 2002), the definition of a Category V protected landscape/seascape is:

> An area of land, with coast and sea as appropriate, where the interaction of people and nature over time has produced an area of distinct character with significant aesthetic, ecological and/or cultural value, and often with high biological diversity.

Safeguarding the integrity of this traditional interaction is vital to the protection and evolution of such an area (IUCN 1994). The protected landscape approach can be particularly appropriate in diverse regions of the world because it:

- Links people's needs with biodiversity conservation.
- Typically comprises a mosaic of land and coastal ownership patterns, including private and communally owned property.
- Can accommodate diverse management regimes, including customary laws governing resource management.
- Has important specific objectives related to conservation of cultural heritage.
- Seeks to bring benefits to local communities and contributes to their wellbeing through the provision of environmental goods and services.
- Has proven to work well in certain indigenous territories where strict protected areas have failed, because it accommodates traditional uses and customary tools for resource management.

The protected landscape and seascape approach engages local communities in stewardship of working landscapes because it:

- Reinforces local responsibility for resource management.
- Builds on existing institutional responsibilities.
- Encourages flexible arrangements for the management of resources, including collaborative management agreements and a range of private land and ocean stewardship tools (Brown 2000).

While Category V appears on lists of protected areas for most countries in the Americas, typically it has been applied to existing designations. Until recently, there has not been a consistent effort to embrace the principles outlined above. A number of recent developments present new opportunities for establishing protected landscapes, as demonstrated by recent progress in Latin America and the Caribbean, as well as in North America. In the USA and Canada, many protected areas have been recognized as meeting the criteria for management as Category V protected landscapes. Sites as diverse as Point Reyes National Seashore, Cuyahoga Valley National Recreation Area, Roosevelt International

Park, and parts of Denali National Park are considered Category V protected landscapes. Most areas are managed by government agencies, though a few are non-traditional in structure, such as the St. Croix Waterway Heritage River. However, exciting opportunities for the protected landscape approach are being created by new models of public–private partnerships. A number of countries are considering reclassification of national parks as a means of addressing conflicts with resident populations.

In Andean South America new sites are being proposed as protected landscapes. At a recent UNESCO-World Heritage Convention meeting held in Arequipa, Peru, 15 cultural landscapes in the Andes were nominated for protection. Among the candidates in Peru are Urabamba (a sacred valley of the Incas) and the Cordillera de Huayhuash in the central sierra of Peru. In Ecuador there is growing interest, at local and national levels, in declaring the Quijos River Valley the country's first protected landscape. Its designation would create a natural corridor among three important protected areas, consolidating them into Ecuador's largest protected area and fostering conservation at an eco-regional scale.

The emergence of networks of private reserves in many countries of Latin America (e.g. Costa Rica, Colombia, Ecuador, Chile and Brazil) is one important way in which the stewardship approach is being applied. While relatively recent in their inception, private reserves are already making significant contributions to conserving cultural and natural heritage, and the movement is gaining momentum. This development holds great potential for protecting working landscapes in the region. A number of countries in the Eastern Caribbean, such as St. Lucia and the British Virgin Islands, are beginning to include Category V protected landscapes in their systems of protected areas. National trusts in these countries have found that the model is highly appropriate for small, intensively settled island countries, where the landscapes reflect human interactions over time, much land is privately owned, and the pattern of ownership necessitates innovative management arrangements.

Challenges

Among the challenges to protecting working landscapes and seascapes is a basic unfamiliarity with designations such as Category V, which is currently not well represented in most national protected area systems. In many countries, particularly in regions such as Latin America and the Caribbean, the complexity of land and ocean use, tenure, and institutional roles can make it hard to work at the scale of landscapes and seascapes. A key challenge lies in coordinating the efforts of diverse actors, all using different mechanisms, to achieve biodiversity conservation goals at the scale of bioregions. The extraordinary growth of private land conservation throughout the Americas and in other regions of the world holds much promise for protecting working landscapes and seascapes. However, there is a need to develop further the criteria and management guidelines for private reserves at a regional level. Legal and institutional mechanisms must be in place to encourage and ensure management agreements. Long-term provision for management and monitoring will be essential to assure adherence to agreements, as well as to evaluate the effectiveness of different approaches.

Coastal resources cannot be used by any interest group as their exclusive right. The sea cannot be regarded as a common basin for effluent disposal. Almost all coastal and marine areas produce or support multiple products and services. Sectoral solutions usually 'transfer' the problem between resources, products and services. Tourism will not flourish if the area loses its attractions to visitors. Fisheries suffer as a result of the pollution problems. Industry and energy facilities can degrade the environment for all other activities. There is, therefore, a need to bring sectoral activities together to achieve a commonly acceptable coastal management framework. The challenge, a recurring theme throughout this book, is to develop the most appropriate institutional arrangement and strategy that will encourage collaboration and consultation at all levels within a responsible ocean governance framework.

Author's concluding remarks

Landscapes and seascapes are dynamic and change along with the communities living in them. As Adrian Phillips of the IUCN World Commission on Protected Areas has observed, this dynamism poses a central dilemma in landscape protection:

> It is not enough therefore to attempt to protect the landscape or seascape as such: attention must be given to the ways of life of those who are architects of the landscape and upon whom the survival of the biodiversity within it depends. [Protection should seek not to] fossilize the ways in which communities use the land [oceans], but rather to encourage sustainable approaches to land use [ocean use] and development.
>
> (Phillips 1997)

Any strategy for protection of working landscapes and seascapes will require tools, adapted to the special characteristics of the local and national context, which can be applied across a mosaic of land ownership and use patterns. It will respect the land and resource rights of indigenous and other traditional peoples. It will rely on approaches that engage local residents and communities, and build on long traditions of caring for natural and cultural heritage.

The stewardship approach offers tremendous potential for sustaining special landscapes and seascapes in diverse regions of the world. The definition of protected landscapes, with its emphasis on the interaction of people and nature over time, implies that people living in the landscape and seascape act as its stewards. In a changing world, new tools are needed to support and ensure stewardship of natural and cultural heritage in keeping with conservation goals.

Stewardship techniques offer great potential to strengthen and extend the impact of conventional protected areas.

Attempts at developing stewardship initiatives should consider:

- *Creating a legal framework conducive to private initiatives:* Incentives (e.g. tax advantages) for conservation and best management practices on private lands must be incorporated into national legislation. Even voluntary and non-binding tools often benefit from governmental recognition. As key actors in stewardship, NGOs require a stable legal basis for establishment and legitimacy as an important sector in civil society.
- *Developing legal and institutional mechanisms to ensure 'conservation in perpetuity':* Long-term provision for management and monitoring will be essential to assure adherence to agreements, as well as to evaluate the effectiveness of different approaches.
- *Creating the climate for productive, enduring partnerships among sectors:* Government agencies charged with protected areas management must have the flexibility to develop appropriate partnerships with NGOs and other private interests. To create an atmosphere of trust and cooperation, government must view these NGOs as true partners, rather than subcontractors; NGOs must be willing to engage in non-adversarial relationships with government; and all parties must be committed to ongoing communication and coordination of efforts.
- *Integrating stewardship into land-use planning and protected areas management:* Private stewardship efforts, however extensive, are no substitute for a strong government role in the planning and protection of natural areas. These efforts should reinforce land-use and ocean-use planning and policy at all levels. At the same time, private initiatives should be viewed not as an afterthought, but as central to meeting protection and management objectives. To this end, coordination among private and public actors is essential.
- *Ensuring participation by all interested parties:* Stewardship relies on public support and participation. Whether through landowner contact or public forums, opportunities must be created for those most affected by land-use and ocean-use decisions to voice their concerns. Value must be placed on local knowledge and traditional resource management systems. Opportunities for collaborative management should be explored.
- *Marshalling the necessary resources:* Funding is necessary for land acquisition and compensation for certain development rights or uses. Often NGOs are in a strong position to raise private funds for these purposes. Fiscal incentives, such as reduced property taxes, may carry a cost in terms of lost revenues to municipalities.
- *Striking a balance between responding to opportunities and taking a strategic approach:* To maintain the ecological integrity of landscapes and seascapes, and protect representative ecosystems, requires strategic approaches. A key challenge lies in coordinating the efforts of diverse actors, all using different

mechanisms, to achieve biodiversity conservation goals at the scale of bioregions. While responding to protection opportunities as they arise, local stewardship initiatives must also be proactive, addressing emerging trends in land use and ocean use, such as re-privatization or increased development pressure. It is unlikely that a structure could serve this purpose which means that a collaborative approach (cluster) should be seriously considered.

In some cases, bilateral relations and mechanisms have been sufficient to address emerging issues. However, in other cases, bilateral relations can act as an impediment to pragmatic measures in addressing challenges that have a multifaceted nature. The increasingly complex transnational nature of maritime challenges in the twenty-first century requires new multilateral and regional approaches. This may mean seeking alternative mechanisms for collective action. Maritime clusters may yet prove to be the best mechanism for this collective action.

References and further reading

Beresford, M. and Phillips, A. (2000). 'Protected landscapes: A conservation model for the 21st century'. *The George Wright Forum*, 17(1): 15–26.

Blue Earth Consultants (2014). 'Ocean stewardship in the private sector'. www.blueearthconsultants.com/wp-content/uploads/2014/07/Blue-Earth-Consultants_Private-Sector_Flyer.pdf (accessed: April 2017).

Brown J. (2000). 'The stewardship approach and its relevance for protected landscapes'. *The George Wright Forum*, 17(1).

Corcoran, E., Nellemann, C., Baker, E., Bos, D., et al. (eds) (2010). *Sick Water? The central role of wastewater management in sutainable development*. Norway, Arendal: United Nations Environment Programme/Grid Arendal. The Crown Estate (2014). 'Marine stewardship program: Annual review 2014'. www.thecrownestate.co.uk/media/389795/marine-stewardship-programme-annual-review-2014.pdf (accessed: April 2017).

IUCN (International Union for Conservation of Nature) (1994). *Guidelines for Protected Area Management Categories*. Gland, Switzerland: ICUN.

Mengerink, K.J., Van Dover, C.L., Ardron, J., Baker, M., et al. (2014). 'A call for deep-ocean stewardship'. *Science*, 344(6185): 696–698.

National Park Service Northeast Region Office of Interpretation and Education (2010). 'Communication strategies that promote ocean stewardship action: A booklet for interpreters and educators'. www.nps.gov/orgs/1412/upload/Ocean-Booklet.pdf (accessed: April 2017).

Organisation for Economic Co-operation and Development (2015). 'The future of the ocean economy: Exploring the prospects for emerging ocean industries to 2030'. OECD workshop: Gothenburg, Sweden, June 2015.

Phillips, A. (1997). 'Landscape approaches to national parks and protected areas' in: J.G. Nelson and R. Serafin (eds), *National Parks and Protected Areas: Keystones to conservation and sustainable development*. Berlin and Heidelberg: Springer-Verlag.

Phillips, A. (2002). 'Management guidelines for IUCN Category V protected areas: Protected landscapes/seascapes'. https://portals.iucn.org/library/sites/library/files/documents/pag-009.pdf (accessed: May 2017).

Rao, A.S. (2014). *Protecting Your Community's Coastal Assets: Local leadership in marine planning*. Vancouver: David Suzuki Foundation.
UNEP (United Nations Environment Program) (2009). *Sustainable Coastal Tourism: An integrated planning and management approach*. Nairobi, Kenya: UNEP.
UNEP (United Nations Environment Program) (2015a). *The Economic Valuation of Wastewater: The cost of action and the cost of no action*. Nairobi, Kenya: UNEP.
UNEP (United Nations Environment Program) (2015b). *Blue Economy: Sharing success stories to inspire change*. Nairobi, Kenya: UNEP.
US National Park Service (n.d.). 'Be a junior ranger.' www.nps.gov/olym/learn/kidsyouth/beajuniorranger.htm (accessed: May 2017).

Index

Note: page numbers in italic type refer to Figures; those in bold type refer to Tables.

For Product Safety Concerns and Information please contact our EU representative GPSR@taylorandfrancis.com
Taylor & Francis Verlag GmbH, Kaufingerstraße 24, 80331 München, Germany

www.ingramcontent.com/pod-product-compliance
Ingram Content Group UK Ltd.
Pitfield, Milton Keynes, MK11 3LW, UK
UKHW010956180926
475701UK00004B/391

* 9 7 8 0 3 6 7 3 3 5 3 6 6 *